Free Study Tips DVD

In addition to the tips and content in this guide, we have created a FREE DVD with helpful study tips to further assist your exam preparation. **This FREE Study Tips DVD provides you with top-notch tips to conquer your exam and reach your goals.**

Our simple request in exchange for the strategy-packed DVD is that you email us your feedback about our study guide. We would love to hear what you thought about the guide, and we welcome any and all feedback—positive, negative, or neutral. It is our #1 goal to provide you with top-quality products and customer service.

To receive your **FREE Study Tips DVD**, email freedvd@apexprep.com. Please put "FREE DVD" in the subject line and put the following in the email:

   a. The name of the study guide you purchased.

   b. Your rating of the study guide on a scale of 1-5, with 5 being the highest score.

   c. Any thoughts or feedback about your study guide.

   d. Your first and last name and your mailing address, so we know where to send your free DVD!

Thank you!

ACT Math Prep Book 2019 & 2020

Math ACT Study Guide 2019 & 2020 with Practice Tests (Includes Two Math Practice Tests)

APEX Test Prep

Table of Contents

Test Taking Strategies...1

FREE DVD OFFER 4

Introduction to the ACT...5

Mathematics Test 7

    Preparing for Higher Mathematics 7

    Integrating Essential Skills..66

    Modeling...79

    Practice Test #1 92

    Answer Explanations #1 107

    Practice Test #2 118

    Answer Explanations #2 134

Test Taking Strategies

1. Reading the Whole Question

A popular assumption in Western culture is the idea that we don't have enough time for anything. We speed while driving to work, we want to read an assignment for class as quickly as possible, or we want the line in the supermarket to dwindle faster. However, speeding through such events robs us from being able to thoroughly appreciate and understand what's happening around us. While taking a timed test, the feeling one might have while reading a question is to find the correct answer as quickly as possible. Although pace is important, don't let it deter you from reading the whole question. Test writers know how to subtly change a test question toward the end in various ways, such as adding a negative or changing focus. If the question has a passage, carefully read the whole passage as well before moving on to the questions. This will help you process the information in the passage rather than worrying about the questions you've just read and where to find them. A thorough understanding of the passage or question is an important way for test takers to be able to succeed on an exam.

2. Examining Every Answer Choice

Let's say we're at the market buying apples. The first apple we see on top of the heap may *look* like the best apple, but if we turn it over we can see bruising on the skin. We must examine several apples before deciding which apple is the best. Finding the correct answer choice is like finding the best apple. Although it's tempting to choose an answer that seems correct at first without reading the others, it's important to read each answer choice thoroughly before making a final decision on the answer. The aim of a test writer might be to get as close as possible to the correct answer, so watch out for subtle words that may indicate an answer is incorrect. Once the correct answer choice is selected, read the question again and the answer in response to make sure all your bases are covered.

3. Eliminating Wrong Answer Choices

Sometimes we become paralyzed when we are confronted with too many choices. Which frozen yogurt flavor is the tastiest? Which pair of shoes look the best with this outfit? What type of car will fill my needs as a consumer? If you are unsure of which answer would be the best to choose, it may help to use process of elimination. We use "filtering" all the time on sites such as eBay® or Craigslist® to eliminate the ads that are not right for us. We can do the same thing on an exam. Process of elimination is crossing out the answer choices we know for sure are wrong and leaving the ones that might be correct. It may help to cover up the incorrect answer choice. Covering incorrect choices is a psychological act that alleviates stress due to the brain being exposed to a smaller amount of information. Choosing between two answer choices is much easier than choosing between all of them, and you have a better chance of selecting the correct answer if you have less to focus on.

4. Sticking to the World of the Question

When we are attempting to answer questions, our minds will often wander away from the question and what it is asking. We begin to see answer choices that are true in the real world instead of true in the world of the question. It may be helpful to think of each test question as its own little world. This world may be different from ours. This world may know as a truth that the chicken came before the egg or may assert that two plus two equals five. Remember that, no matter what hypothetical nonsense may be in the question, assume it to be true. If the question states that the chicken came before the egg, then choose your answer based on that truth. Sticking to the world of the question means placing all of our biases and assumptions aside and relying on the question to guide us to the correct answer. If we are simply looking for answers that are correct based on our own judgment, then we may choose incorrectly. Remember an answer that is true does not necessarily answer the question.

5. Key Words

If you come across a complex test question that you have to read over and over again, try pulling out some key words from the question in order to understand what exactly it is asking. Key words may be words that surround the question, such as *main idea, analogous, parallel, resembles, structured,* or *defines.* The question may be asking for the main idea, or it may be asking you to define something. Deconstructing the sentence may also be helpful in making the question simpler before trying to answer it. This means taking the sentence apart and obtaining meaning in pieces, or separating the question from the foundation of the question. For example, let's look at this question:

> Given the author's description of the content of paleontology in the first paragraph, which of the following is most parallel to what it taught?

The question asks which one of the answers most *parallels* the following information: The *description* of paleontology in the first paragraph. The first step would be to see *how* paleontology is described in the first paragraph. Then, we would find an answer choice that parallels that description. The question seems complex at first, but after we deconstruct it, the answer becomes much more attainable.

6. Subtle Negatives

Negative words in question stems will be words such as *not, but, neither,* or *except.* Test writers often use these words in order to trick unsuspecting test takers into selecting the wrong answer—or, at least, to test their reading comprehension of the question. Many exams will feature the negative words in all caps (*which of the following is NOT an example*), but some questions will add the negative word seamlessly into the sentence. The following is an example of a subtle negative used in a question stem:

> According to the passage, which of the following is *not* considered to be an example of paleontology?

If we rush through the exam, we might skip that tiny word, *not,* inside the question, and choose an answer that is opposite of the correct choice. Again, it's important to read the question fully, and double check for any words that may negate the statement in any way.

7. Spotting the Hedges

The word "hedging" refers to language that remains vague or avoids absolute terminology. Absolute terminology consists of words like *always, never, all, every, just, only, none,* and *must.* Hedging refers to words like *seem, tend, might, most, some, sometimes, perhaps, possibly, probability,* and *often.* In some cases, we want to choose answer choices that use hedging and avoid answer choices that use absolute terminology. It's important to pay attention to what subject you are on and adjust your response accordingly.

8. Restating to Understand

Every now and then we come across questions that we don't understand. The language may be too complex, or the question is structured in a way that is meant to confuse the test taker. When you come across a question like this, it may be worth your time to rewrite or restate the question in your own words in order to understand it better. For example, let's look at the following complicated question:

> Which of the following words, if substituted for the word *parochial* in the first paragraph, would LEAST change the meaning of the sentence?

Let's restate the question in order to understand it better. We know that they want the word *parochial* replaced. We also know that this new word would "least" or "not" change the meaning of the sentence. Now let's try the sentence again:

> Which word could we replace with *parochial,* and it would not change the meaning?

Restating it this way, we see that the question is asking for a synonym. Now, let's restate the question so we can answer it better:

> Which word is a synonym for the word *parochial?*

Before we even look at the answer choices, we have a simpler, restated version of a complicated question.

9. Predicting the Answer

After you read the question, try predicting the answer *before* reading the answer choices. By formulating an answer in your mind, you will be less likely to be distracted by any wrong answer choices. Using predictions will also help you feel more confident in the answer choice you select. Once you've chosen your answer, go back and reread the question and answer choices to make sure you have the best fit. If you have no idea what the answer may be for a particular question, forego using this strategy.

10. Avoiding Patterns

One popular myth in grade school relating to standardized testing is that test writers will often put multiple-choice answers in patterns. A runoff example of this kind of thinking is that the most common answer choice is "C," with "B" following close behind. Or, some will advocate certain made-up word patterns that simply do not exist. Test writers do not arrange their correct answer choices in any kind of pattern; their choices are randomized. There may even be times where the correct answer choice will be the same letter for two or three questions in a row, but we have no way of knowing when or if this might happen. Instead of trying to figure out what choice the test writer probably set as being correct, focus on what the *best answer choice* would be out of the answers you are presented with. Use the tips above, general knowledge, and reading comprehension skills in order to best answer the question, rather than looking for patterns that do not exist.

its ok to have

a a a...

3

FREE DVD OFFER

Achieving a high score on your exam depends not only on understanding the content, but also on understanding how to apply your knowledge and your command of test taking strategies. **Because your success is our primary goal, we offer a FREE Study Tips DVD. It provides top-notch test taking strategies to help you optimize your testing experience.**

Our simple request in exchange for the strategy-packed DVD is that you email us your feedback about our study guide.

To receive your **FREE Study Tips DVD**, email freedvd@apexprep.com. Please put "FREE DVD" in the subject line and put the following in the email:

a. The name of the study guide you purchased.

b. Your rating of the study guide on a scale of 1-5, with 5 being the highest score.

c. Any thoughts or feedback about your study guide.

d. Your first and last name and your mailing address, so we know where to send your free DVD!

Introduction to the ACT

Function of the Test

The ACT is one of the two national standardized college entrance examinations, with the SAT serving as the other option. Most hopeful college-bound students take the ACT, SAT, or both. For admissions purposes, every four-year college and university in the United States accepts ACE scores, and some schools require it. More so than the SAT, which primarily serves as an aptitude test, the ACT is often used for course placement purposes because it measures academic achievement on content addressed in high school classes. Twelve states also require that all high school juniors in the state take the ACT, and eight additional states have counties that require the exam.

Most ACT test takers are prospective college students who are currently in their junior or senior year of high school. More than 2 millions students graduating in the class of 2017 took the ACT.

Test Administration

The ACT is offered on seven dates in the U.S. and Canada, and on six dates internationally throughout the year. The exam is usually administered at high schools or colleges, but other locations may be offered. The registration fee includes the cost to submit score reports to four colleges, but for an additional fee, students can send scores to additional institutions. There is a separate registration fee incurred for the optional writing section. Some high schools cover the fees for their students, so prospective test takers are advised to contact the guidance counselor at their school.

Test takers can retake the ACT every time the test is offered, up to a maximum of twelve times. However, different colleges and universities sometimes have limits on the number of retakes they will consider. Moreover, scores from the various test sections cannot be combined from different test attempts. Reasonable accommodations will be provided to test takers with appropriate documentation for a variety of disabilities.

Test Format

Test takers are given a total of 175 minutes to complete the 215 multiple-choice questions in four subject subtests (English, Mathematics, Reading, and Science) of the ACT. It also has an optional Writing Test, which involves writing an essay, which takes an additional forty minutes. Some colleges and universities require the essay for admission.

The English Test consists of 75 questions that address the production of writing, knowledge of language, and conventions of standard English. The 60-question Mathematics Test involves number sense, algebra, functions, geometry, statistics and probability, and modeling. Calculators that meet certain calculator requirements are permitted. The Reading Test is contains four written passages, with ten questions per passage addressing comprehension skills, the ability to make inferences and draw conclusions, and apply and integrate knowledge. The Science Test contains 40 questions that require interpreting data, understanding scientific investigations, and evaluating models and results.

The Writing Test is always given at the end of the exam so that test takers opting not to take it may leave after completing the other four subtests. This section consists of one essay in which students must analyze three different perspectives on a broad social issue and reconcile them in a cohesive essay. The following chart provides the breakdown of the sections of the ACT:

Subtest	Length	Number of Questions
English	45 minutes	75
Mathematics	60 minutes	60
Reading	35 minutes	40
Science	35 minutes	40
Writing (optional)	40 minutes	1 essay

Scoring

Score reports are typically available two weeks after the date of administration. Because there is no penalty for incorrect answers, test takers are encouraged to answer every question, even if they have to guess. For each of the four required subtests, test takers receive a score between 1 and 36. These scores are then averaged together to yield a Composite Score, which is the primary score reported as an "ACT score." The most prestigious colleges and universities are typically looking for Composite Scores greater than 30 in order to consider an applicant for admissions. Other selective schools typically expect candidates to have scores just under 30. Average institutions are more likely to set the bar lower (perhaps in the low 20s), while community colleges usually accept students with scores in the high teens. In 2016 and 2017, the mean Composite Score among all test takers (including those not applying to college) was 20.9.

The Writing Test is scored on a scale that ranges from 2 to 12 scale. In 2016 and 2017, the mean score was 6.7.

Mathematics Test

Preparing for Higher Mathematics

Number & Quantity

Real and Complex Number Systems

Whole numbers are the numbers 0, 1, 2, 3, ... Examples of other whole numbers would be 413 and 8,431. Notice that numbers such as 4.13 and $\frac{1}{4}$ are not included in whole numbers. **Counting numbers**, also known as **natural numbers**, consist of all whole numbers except for the zero. In set notation, the natural numbers are the set $\{1, 2, 3, ... \}$. The entire set of whole numbers and negative versions of those same numbers comprise the set of numbers known as **integers.** Therefore, in set notation, the integers are $\{..., -3, -2, -1, 0, 1, 2, 3, ... \}$. Examples of other integers are $-4,981$ and $90,131$. A number line is a great way to visualize the integers. Integers are labeled on the following number line:

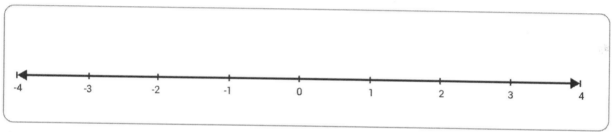

The arrows on the right- and left-hand sides of the number line show that the line continues indefinitely in both directions.

Fractions also exist on the number line and are considered parts of a whole. For example, if an entire pie is cut into two pieces, each piece is half of the pie, or $\frac{1}{2}$. The top number in any fraction, known as the **numerator,** defines how many parts there are. The bottom number, known as the **denominator,** states how many pieces the whole is divided into. Fractions can also be negative or written in their corresponding decimal form.

A **decimal** is a number that uses a decimal point and numbers to the right of the decimal point representing the part of the number that is less than 1. For example, 3.5 is a decimal and is equivalent to the fraction $\frac{7}{2}$ or the mixed number $3\frac{1}{2}$. The decimal is found by dividing 2 into 7. Other examples of fractions are $\frac{2}{7}, \frac{-3}{14}$, and $\frac{14}{27}$.

Any number that can be expressed as a fraction is known as a **rational number.** Basically, if a and b are any integers and $b \neq 0$, then $\frac{a}{b}$ is a rational number. Any integer can be written as a fraction where the denominator is 1; therefore, the set of rational numbers consist of all fractions and all integers.

Any number that is not rational is known as an **irrational number.** Consider the number $\pi = 3.141592654 ....$ The decimal portion of that number extends indefinitely. In that situation, the number can never be written as a fraction. Another example of an irrational number is $\sqrt{2} = 1.414213662 ....$ Again, this number cannot be written as a ratio of two integers.

7

Together, the set of all rational and irrational numbers makes up the **real numbers.** The number line contains all real numbers. To graph a number other than an integer on a number line, the number will need to be plotted between two integers. For example, 3.5 would be plotted halfway between 3 and 4.

Even numbers are integers that are divisible by 2. For example, 6, 100, 0, and −200 are all even numbers. **Odd numbers** are integers that are not divisible by 2. If an odd number is divided by 2, the result is a fraction. For example, −5, 11, and −121 are odd numbers.

Prime numbers consist of natural numbers greater than 1 that are not divisible by any other natural numbers other than themselves and 1. For example, 3, 5, and 7 are prime numbers. If a natural number is not prime, it is known as a **composite number**. 8 is a composite number because it is divisible by both 2 and 4, which are natural numbers other than itself and 1.

The **absolute value** of any real number is the distance from that number to 0 on the number line. The absolute value of a number can never be negative. For example, the absolute value of both 8 and −8 is 8 because they are both 8 units away from 0 on the number line. This is written as $|8| = |-8| = 8$.

Performing Arithmetic Operations with Rational Numbers
The four basic operations include addition, subtraction, multiplication, and division. The result of addition is a **sum**, the result of subtraction is a **difference**, the result of multiplication is a **product**, and the result of division is a **quotient**. Each type of operation can be used when working with rational numbers; however, the basic operations need to be understood first while using simpler numbers before working with fractions and decimals.

Performing these operations should first be learned using whole numbers. Addition needs to be done column by column. To add two whole numbers, add the ones column first, then the tens columns, then the hundreds, etc. If the sum of any column is greater than 9, a one must be carried over to the next column. For example, the following is the result of 482 + 924:

$$
\begin{array}{r}
{}^{1} \\
482 \\
+924 \\
\hline
1406
\end{array}
$$

Notice that the sum of the tens column was 10, so a one was carried over to the hundreds column. Subtraction is also performed column by column. Subtraction is performed in the ones column first, then the tens, etc. If the number on top is smaller than the number below, a one must be borrowed from the column to the left. For example, the following is the result of 5,424 − 756:

$$
\begin{array}{r}
4\ 13\ 11\ 14 \\
\cancel{5}\ \cancel{4}\ \cancel{2}\ \cancel{4} \\
-\ 7\ 5\ 6 \\
\hline
4\ 6\ 6\ 8
\end{array}
$$

Notice that a one is borrowed from the tens, hundreds, and thousands place. After subtraction, the answer can be checked through addition. A check of this problem would be to show that 756 + 4,668 = 5,424.

Multiplication of two whole numbers is performed by writing one on top of the other. The number on top is known as the **multiplicand,** and the number below is the **multiplier**. Perform the multiplication by multiplying the multiplicand by each digit of the multiplier. Make sure to place the ones value of each result under the multiplying digit in the multiplier. Each value to the right is then a 0. The product is found by adding each product. For example, the following is the process of multiplying 46 times 37 where 46 is the multiplicand and 37 is the multiplier:

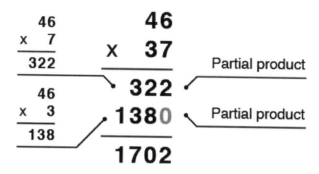

Finally, division can be performed using long division. When dividing a number by another number, the first number is known as the **dividend,** and the second is the **divisor.** For example, with $a \div b = c$, a is the dividend, b is the divisor, and c is the quotient. For long division, place the dividend within the division symbol and the divisor on the outside. For example, with $8,764 \div 4$, refer to the first problem in the diagram below. First, there are 2 4's in the first digit, 8. This number 2 gets written above the 8. Then, multiply 4 times 2 to get 8, and that product goes below the 8. Subtract to get 8, and then carry down the 7. Continue the same steps. $7 \div 4 = 1$ R3, so 1 is written above the 7. Multiply 4 times 1 to get 4, and write it below the 7. Subtract to get 3, and carry the 6 down next to the 3. Resulting steps give a 9 and a 1. The final subtraction results in a 0, which means that 8,764 is divisible by 4. There are no remaining numbers.

The second example shows that $4,536 \div 216 = 21$. The steps are a little different because 216 cannot be contained in 4 or 5, so the first step is placing a 2 above the 3 because there are 2 216's in 453. Finally, the third example shows that $546 \div 31 = 17$ R19. The 19 is a remainder. Notice that the final subtraction does not result in a 0, which means that 546 is not divisible by 31.

The remainder can also be written as a fraction over the divisor to say that:

$$546 \div 31 = 17\frac{19}{31}$$

```
  2 1 9 1              2 1            1 7  r  1 9
4 8 7 6 4        2 1 6 4 5 3 6     3 1 5 4 6
  8                    4 3 2            3 1
  0 7                  2 1 6            2 3 6
    4                  2 1 6            2 1 7
    3 6                    0              1 9
    3 6
      0 4
        4
        0
```

If a division problem relates to a real-world application, and a remainder does exist, it can have meaning. For example, consider the third example, $546 \div 31 = 17R19$. Let's say that we had $546 to spend on calculators that cost $31 each, and we wanted to know how many we could buy. The division problem would answer this question. The result states that 17 calculators could be purchased, with $19 left over. Notice that the remainder will never be greater than or equal to the divisor.

Once the operations are understood with whole numbers, they can be used with integers. There are many rules surrounding operations with negative numbers. First, consider addition with integers. The sum of two numbers can first be shown using a number line. For example, to add $-5 + (-6)$, plot the point -5 on the number line. Then, because a negative number is being added, move 6 units to the left. This process results in landing on -11 on the number line, which is the sum of -5 and -6. If adding a positive number, move to the right.

Visualizing this process using a number line is useful for understanding; however, it is not efficient. A quicker process is to learn the rules. When adding two numbers with the same sign, add the absolute values of both numbers, and use the common sign of both numbers as the sign of the sum. For example, to add $-5 + (-6)$, add their absolute values $5 + 6 = 11$. Then, introduce a negative number because both addends are negative. The result is -11. To add two integers with unlike signs, subtract the lesser absolute value from the greater absolute value, and apply the sign of the number with the greater absolute value to the result. For example, the sum $-7 + 4$ can be computed by finding the difference $7 - 4 = 3$ and then applying a negative because the value with the larger absolute value is negative. The result is -3. Similarly, the sum $-4 + 7$ can be found by computing the same difference but leaving it as a positive result because the addend with the larger absolute value is positive. Also, recall that any number plus 0 equals that number. This is known as the **Addition Property of 0.**

Subtracting two integers can be computed by changing to addition to avoid confusion. The rule is t the first number to the opposite of the second number. The opposite of a number is the number on the other side of 0 on the number line, which is the same number of units away from 0. For example, -2 and 2 are opposites. Consider $4 - 8$. Change this to adding the opposite as follows: $4 + (-8)$. Then, follow the rules of addition of integers to obtain -4. Secondly, consider $-8 - (-2)$. Change this problem to adding the opposite as $-8 + 2$, which equals -6. Notice that subtracting a negative number is really adding a positive number.

Multiplication and division of integers are actually less confusing than addition and subtraction because the rules are simpler to understand. If two factors in a multiplication problem have the same sign, the result is positive. If one factor is positive and one factor is negative, the result, known as the **product,** is negative. For example, $(-9)(-3) = 27$ and $9(-3) = -27$. Also, a number times 0 always results in 0. If a problem consists of more than a single multiplication, the result is negative if it contains an odd number of negative factors, and the result is positive if it contains an even number of negative factors. For example, $(-1)(-1)(-1)(-1) = 1$ and $(-1)(-1)(-1)(-1)(-1) = 1$. These two examples of multiplication also bring up another concept. Both are examples of repeated multiplication, which can be written in a more compact notation using exponents. The first example can be written as $(-1)^4 = 1$, and the second example can be written as $(-1)^5 = -1$. Both are exponential expressions, -1 is the base in both instances, and 4 and 5 are the respective exponents. Note that a negative number raised to an odd power is always negative, and a negative number raised to an even power is always positive. Also, $(-1)^4$ is not the same as -1^4. In the first expression, the negative is included in the parentheses, but it is not in the second expression. The second expression is found by evaluating 1^4 first to get 1 and then by applying the negative sign to obtain -1.

A similar theory applies within division. First, consider some vocabulary. When dividing 14 by 2, it can be written in the following ways: $14 \div 2 = 7$ or $\frac{14}{2} = 7$. 14 is the **dividend,** 2 is the **divisor**, and 7 is the **quotient.** If two numbers in a division problem have the same sign, the quotient is positive. If two numbers in a division problem have different signs, the quotient is negative. For example, $14 \div (-2) = -7$, and $-14 \div (-2) = 7$. To check division, multiply the quotient by the divisor to obtain the dividend. Also, remember that 0 divided by any number is equal to 0. However, any number divided by 0 is undefined. It just does not make sense to divide a number by 0 parts.

If more than one operation is to be completed in a problem, follow the Order of Operations. The mnemonic device, PEMDAS, for the order of operations states the order in which addition, subtraction, multiplication, and division needs to be done. It also includes when to evaluate operations within grouping symbols and when to incorporate exponents. PEMDAS, which some remember by thinking "please excuse my dear Aunt Sally," refers to parentheses, exponents, multiplication, division, addition, and subtraction. First, within an expression, complete any operation that is within parentheses, or any other grouping symbol like brackets, braces, or absolute value symbols. Note that this does not refer to the case when parentheses are used to represent multiplication like $(2)(5)$ because in such cases, an operation is not within parentheses like it is in $(2 \cdot 5)$. Then, any exponents must be computed. Next, multiplication and division are performed from left to right. Finally, addition and subtraction are performed from left to right.

The following is an example in which the operations within the parentheses need to be performed first, so the order of operations must be applied to the exponent, subtraction, addition, and multiplication within the grouping symbol:

$$9 - 3(3^2 - 3 + 4 \cdot 3)$$

$$9 - 3(3^2 - 3 + 4 \cdot 3)$$ Work within the parentheses first

$$= 9 - 3(9 - 3 + 12)$$

$$= 9 - 3(18)$$

$$= 9 - 54$$

$$= -45$$

Once the rules for integers are understood, move on to learning how to perform operations with fractions and decimals. Recall that a rational number can be written as a fraction and can be converted to a decimal through division. If a rational number is negative, the rules for adding, subtracting, multiplying, and dividing integers must be used. If a rational number is in fraction form, performing addition, subtraction, multiplication, and division is more complicated than when working with integers. First, consider addition. To add two fractions having the same denominator, add the numerators and then reduce the fraction. When an answer is a fraction, it should always be in lowest terms. **Lowest terms** means that every common factor, other than 1, between the numerator and denominator is divided out. For example:

$$\frac{2}{8} + \frac{4}{8} = \frac{6}{8}$$

$$\frac{6 \div 2}{8 \div 2} = \frac{3}{4}$$

Both the numerator and denominator of $\frac{6}{8}$ have a common factor of 2, so 2 is divided out of each number to put the fraction in lowest terms. If denominators are different in an addition problem, the fractions must be converted to have common denominators. The **least common denominator (LCD)** of all the given denominators must be found, and this value is equal to the **least common multiple (LCM)** of the denominators. This non-zero value is the smallest number that is a multiple of both denominators. Then, rewrite each original fraction as an equivalent fraction using the new denominator. Once in this form, apply the process of adding with like denominators. For example, consider $\frac{1}{3} + \frac{4}{9}$. The LCD is 9 because it is the smallest multiple of both 3 and 9. The fraction $\frac{1}{3}$ must be rewritten with 9 as its denominator. Therefore, multiply both the numerator and denominator by 3. Multiplying by $\frac{3}{3}$ is the same as multiplying by 1, which does not change the value of the fraction. Therefore, an equivalent fraction is $\frac{3}{9}$, and $\frac{1}{3} + \frac{4}{9} = \frac{3}{9} + \frac{4}{9} = \frac{7}{9}$, which is in lowest terms. Subtraction is performed in a similar manner; once the denominators are equal, the numerators are then subtracted.

The following is an example of addition of a positive and a negative fraction:

$$-\frac{5}{12}+\frac{5}{9}$$

$$-\frac{5\times3}{12\times3}+\frac{5\times4}{9\times4}$$

$$-\frac{15}{36}+\frac{20}{36}=\frac{5}{36}$$

Common denominators are not used in multiplication and division. To multiply two fractions, multiply the numerators together and the denominators together. Then, write the result in lowest terms. For example:

$$\frac{2}{3}\times\frac{9}{4}=\frac{18}{12}=\frac{3}{2}$$

Alternatively, the fractions could be factored first to cancel out any common factors before performing the multiplication. For example:

$$\frac{2}{3}\times\frac{9}{4}=\frac{2}{3}\times\frac{3\times3}{2\times2}=\frac{3}{2}$$

This second approach is helpful when working with larger numbers, as common factors might not be obvious. Multiplication and division of fractions are related because the division of two fractions is changed into a multiplication problem. Division of a fraction is equivalent to multiplication of the **reciprocal** of the second fraction, so that second fraction must be inverted, or "flipped," to be in reciprocal form. For example:

$$\frac{11}{15}\div\frac{3}{5}=\frac{11}{15}\times\frac{5}{3}=\frac{55}{45}=\frac{11}{9}$$

The fraction $\frac{5}{3}$ is the reciprocal of $\frac{3}{5}$.

It is possible to multiply and divide numbers containing a mix of integers and fractions. In this case, convert the integer to a fraction by placing it over a denominator of 1. For example, a division problem involving an integer and a fraction is:

$$3\div\frac{1}{2}$$

$$\frac{3}{1}\times\frac{2}{1}$$

$$\frac{6}{1}=6$$

Finally, when performing operations with rational numbers that are negative, the same rules apply as when performing operations with integers. For example, a negative fraction multiplied by a negative fraction results in a positive value, and a negative fraction subtracted from a negative fraction results in a negative value.

Operations can be performed on rational numbers in decimal form. Recall that to write a fraction as an equivalent decimal expression, divide the numerator by the denominator.

For example:

$$\frac{1}{8} = 1 \div 8 = 0.125$$

With the case of decimals, it is important to keep track of place value. To add decimals, make sure the decimal places are in alignment so that the numbers are lined up with their decimal points and add vertically. If the numbers do not line up because there are extra or missing place values in one of the numbers, then zeros may be used as placeholders. For example, $0.123 + 0.23$ becomes:

$$\begin{array}{r} 0.123 \\ + \ 0.230 \\ \hline 0.353 \end{array}$$

Subtraction is done the same way. Multiplication and division are more complicated. To multiply two decimals, place one on top of the other as in a regular multiplication process and do not worry about lining up the decimal points. Then, multiply as with whole numbers, ignoring the decimals. Finally, in the solution, insert the decimal point as many places to the left as there are total decimal values in the original problem. Here is an example of a decimal multiplication:

$$\begin{array}{r} 0.52 \quad \textit{2 decimal places} \\ \times \quad 0.2 \quad \textit{1 decimal place} \\ \hline 0.104 \quad \textit{3 decimal places} \end{array}$$

The answer to 52 times 2 is 104, and because there are three decimal values in the problem, the decimal point is positioned three units to the left in the answer.

The decimal point plays an integral role throughout the whole problem when dividing with decimals. First, set up the problem in a long division format. If the divisor is not an integer, the decimal must be moved to the right as many units as needed to make it an integer. The decimal in the dividend must be moved to the right the same number of places to maintain equality.

Then, division is completed normally. Here is an example of long division with decimals:

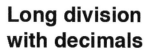

**Long division
with decimals**

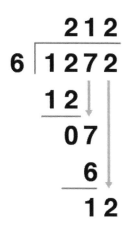

```
        2 1 2
    6 | 1 2 7 2
        1 2 ↓
        0 7
          6 ↓
        1 2
```

Because the decimal point is moved two units to the right in the divisor of 0.06 to turn it into the integer 6, it is also moved two units to the right in the dividend of 12.72 to make it 1,272. The result is 212, and remember that a division problem can always be checked by multiplying the answer by the divisor to see if the result is equal to the dividend.

Sometimes it is helpful to round answers that are in decimal form. First, find the place to which the rounding needs to be done. Then, look at the digit to the right of it. If that digit is 4 or less, the number in the place value to its left stays the same, and everything to its right becomes a 0. This process is known as rounding down. If that digit is 5 or higher, round up by increasing the place value to its left by 1, and every number to its right becomes a 0. If those 0's are in decimals, they can be dropped. For example, 0.145 rounded to the nearest hundredth place would be rounded up to 0.15, and 0.145 rounded to the nearest tenth place would be rounded down to 0.1.

Another operation that can be performed on rational numbers is the square root. Dealing with real numbers only, the **positive square root** of a number is equal to one of the two repeated positive factors of that number. For example:

$$\sqrt{49} = \sqrt{7 \cdot 7} = 7$$

A **perfect square** is a number that has a whole number as its square root. Examples of perfect squares are 1, 4, 9, 16, 25, etc. If a number is not a perfect square, an approximation can be used with a calculator. For example, $\sqrt{67} = 8.185$, rounded to the nearest thousandth place. The square root of a fraction involving perfect squares involves breaking up the problem into the square root of the numerator separate from the square root of the denominator. For example:

$$\sqrt{\frac{16}{25}} = \frac{\sqrt{16}}{\sqrt{25}} = \frac{4}{5}$$

15

If the fraction does not contain perfect squares, a calculator can be used. Therefore, $\sqrt{\frac{2}{5}} = 0.632$, rounded to the nearest thousandth place. A common application of square roots involves the Pythagorean theorem. Given a right triangle, the sum of the squares of the two legs equals the square of the hypotenuse.

For example, consider the following right triangle:

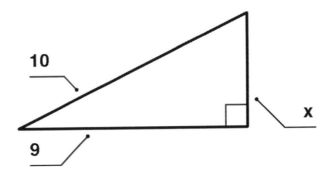

The missing side, x, can be found using the Pythagorean theorem.

$$9^2 + x^2 = 10^2$$

$$81 + x^2 = 100$$

$$x^2 = 19$$

To solve for x, take the square root of both sides. Therefore, $x = \sqrt{19} = 4.36$, which has been rounded to two decimal places.

In addition to the square root, the cube root is another operation. If a number is a **perfect cube**, the cube root of that number is equal to one of the three repeated factors. For example:

$$\sqrt[3]{27} = \sqrt[3]{3 \cdot 3 \cdot 3} = 3$$

Also, unlike square roots, a negative number has a cube root. The result is a negative number. For example:

$$\sqrt[3]{-27}$$

$$\sqrt[3]{(-3)(-3)(-3)} = -3$$

Similar to square roots, if the number is not a perfect cube, a calculator can be used to find an approximation. Therefore, $\sqrt[3]{\frac{2}{3}} = 0.873$, rounded to the nearest thousandth place.

Higher-order roots also exist. The number relating to the root is known as the **index.** Given the following root, $\sqrt[3]{64}$, 3 is the index, and 64 is the **radicand.** The entire expression is known as the **radical.** Higher-order roots exist when the index is larger than 3. They can be broken up into two groups: even and odd roots. **Even roots**, when the index is an even number, follow the properties of square roots. A negative number does not have an even root, and an even root is found by finding the single factor that is repeated the same number of times as the index in the radicand.

For example, the fifth root of 32 is equal to 2 because:

$$\sqrt[5]{32} = \sqrt[5]{2 \cdot 2 \cdot 2 \cdot 2 \cdot 2} = 2$$

Odd roots, when the index is an odd number, follow the properties of cube roots. A negative number has an odd root. Similarly, an odd root is found by finding the single factor that is repeated that many times to obtain the radicand. For example, the 4th root of 81 is equal to 3 because $3^4 = 81$. This radical is written as $\sqrt[4]{81} = 4$. Higher-order roots can also be evaluated on fractions and decimals, for example, because:

$$\left(\frac{2}{7}\right)^4 = \frac{16}{2{,}401}, \sqrt[4]{\frac{16}{2{,}401}} = \frac{2}{7}$$

and because:

$$(0.1)^5 = 0.00001, \sqrt[5]{0.00001} = 0.1$$

Sometimes, when performing operations in rational numbers, it might be helpful to round the numbers in the original problem to get a rough estimate of what the answer should be. For example, if you walked into a grocery store and had a $20 bill, your approach might be to round each item to the nearest dollar and add up all the items to make sure that you will have enough money when you check out. This process involves obtaining an estimation of what the exact total would be. In other situations, it might be helpful to round to the nearest $10 amount or $100 amount. **Front-end rounding** might be helpful as well in many situations. In this type of rounding, each number is rounded to the highest possible place value. Therefore, all digits except the first digit become 0. Consider a situation in which you are at the furniture store and want to estimate your total on three pieces of furniture that cost $434.99, $678.99, and $129.99. Front-end rounding would round these three amounts to $400, $700, and $100. Therefore, the estimate of your total would be $400 + $700 + $100 = $1,200, compared to the exact total of $1,243.97. In this situation, the estimate is not that far off the exact answer.

Rounding is useful in both approximating an answer when an exact answer is not needed and for comparison when an exact answer is needed. For instance, if you had a complicated set of operations to complete and your estimate was $1,000, if you obtained an calculated answer of $100,000, something is off. You might want to check your work to see if a mistake was made because an estimate should not be that different from an exact answer. Estimates can also be helpful with square roots. If a square root of a number is not known, the closest perfect square can be found for an approximation. For example, $\sqrt{50}$ is not equal to a whole number, but 50 is close to 49, which is a perfect square, and $\sqrt{49} = 7$. Therefore, $\sqrt{50}$ is a little bit larger than 7. The actual approximation, rounded to the nearest thousandth, is 7.071.

Ordering and Comparing Rational Numbers
Ordering rational numbers is a way to compare two or more different numerical values. Determining whether two amounts are equal, or if one is less than or greater than the other is the basis for comparing both positive and negative numbers. Also, a group of numbers can be compared by ordering them from the smallest value to the largest value. A few symbols are necessary to use when ordering rational numbers. The equals sign, =, shows that the two quantities on either side of the symbol have the same value. For example, $\frac{12}{3} = 4$ because both values are equivalent. Another symbol that is used to compare numbers is <, which represents "less than." With this symbol, the smaller number is placed on the left and the larger number is placed on the right. Always remember that the symbol's "mouth" opens up to the larger number. When comparing negative and positive numbers, it is important to remember that the

number occurring to the left on the number line is always smaller and is placed to the left of the symbol. This idea might seem confusing because some values could appear at first glance to be larger, even though they are not. For example, $-5 < 4$ is read "negative 5 is less than 4." Here is an image of a number line to help visualize relationships:

The symbol $\leq$ represents "less than or equal to," and it joins $<$ with equality. Therefore, both $-5 \leq 4$ and $-5 \leq -5$ are true statements and "-5 is less than or equal to both 4 and -5." Other symbols are $>$ and $\geq$, which represent "greater than" and "greater than or equal to." Both $4 \geq -1$ and $-1 \geq -1$ are correct ways to use these symbols.

Here is a chart of these four inequality symbols:

Symbol	Definition
$<$	less than
$\leq$	less than or equal to
$>$	greater than
$\geq$	greater than or equal to

Comparing integers is a straightforward process, especially when using the number line, but the comparison of decimals and fractions is not as obvious. When comparing two non-negative decimals, compare digit by digit, starting from the left. The larger value contains the first larger digit. For example, 0.1456 is larger than 0.1234 because the value 4 in the hundredths place in the first decimal is larger than the value 2 in the hundredths place in the second decimal. When comparing a fraction with a decimal, convert the fraction to a decimal and then compare in the same manner. Finally, there are a few options when comparing multiple fractions. If two non-negative fractions have the same denominator, the fraction with the larger numerator is the larger value. If they have the same numerator but different denominators (such as $\frac{3}{5}$ and $\frac{3}{10}$), the fraction whose denominator is a smaller number (in this case, the $\frac{3}{5}$ because 5 is less than 10) is larger.

If the two fractions have different numerators and denominators, they can be converted to equivalent fractions with a common denominator to be compared, or they can be converted to decimals to be compared. When comparing two negative decimals or fractions, a different approach must be used. It is important to remember that the smaller number exists to the left on the number line. Therefore, when comparing two negative decimals by place value, the number with the larger first place value is smaller due to the negative sign. Whichever value is closer to 0 is larger. For instance, -0.456 is larger than -0.498 because of the values in the hundredth places. If two negative fractions have the same denominator, the fraction with the larger numerator is smaller because of the negative sign.

Solving Real-World One- or Multi-Step Problems with Rational Numbers
One-step problems take only one mathematical step to solve. For example, solving the equation $5x = 45$ is a one-step problem because the one step of dividing both sides of the equation by 5 is the only step necessary to obtain the solution $x = 9$. The **multiplication principle of equality** is the one step used to isolate the variable. The equation is of the form $ax = b$, where a and b are rational numbers. Similarly, the

addition principle of equality could be the one step needed to solve a problem. In this case, the equation would be of the form $x + a = b$ or $x - a = b$, for real numbers a and b.

A multi-step problem requires more than one step to find the solution, or it might consist of solving more than one equation. An equation that involves both the addition principle and the multiplication principle is a two-step problem, and an example of such an equation is:

$$2x - 4 = 5$$

Solving involves adding 4 to both sides and then dividing both sides by 2. An example of a two-step problem involving two separate equations is:

$$y = 3x, \qquad 2x + y = 4$$

The two equations form a system of two equations that must be solved together in two variables. The system can be solved by the substitution method. Since y is already solved for in terms of x, plug $3x$ in for y into the equation $2x + y = 4$, resulting in $2x + 3x = 4$. Therefore:

$$5x = 4 \text{ and } x = \frac{4}{5}$$

Because there are two variables, the solution consists of both a value for x and for y. Substitute $x = \frac{4}{5}$ into either original equation to find y. The easiest choice is $y = 3x$. Therefore:

$$y = 3 \times \frac{4}{5} = \frac{12}{5}$$

The solution can be written as the ordered pair $\left(\frac{4}{5}, \frac{12}{5}\right)$.

Real-world problems can be translated into both one-step and multi-step problems. In either case, the word problem must be translated from the verbal form into mathematical expressions and equations that can be solved using algebra. An example of a one-step real-world problem is the following: A cat weighs half as much as a dog living in the same house. If the dog weighs 14.5 pounds, how much does the cat weigh? To solve this problem, an equation can be used.

In any word problem, the first step is to define variables that represent the unknown quantities. For this problem, let x be equal to the unknown weight of the cat. Because two times the weight of the cat equals 14.5 pounds, the equation to be solved is: $2x = 14.5$. Use the multiplication principle to divide both sides by 2. Therefore, $x = 7.25$. The cat weighs 7.25 pounds.

Most of the time, real-world problems are more difficult than this one and consist of multi-step problems. The following is an example of a multi-step problem: The sum of two consecutive page numbers is equal to 437. What are those page numbers? First, define the unknown quantities. If x is equal to the first page number, then $x + 1$ is equal to the next page number because they are consecutive integers. Their sum is equal to 437, and this statement translates to the equation $x + x + 1 = 437$. To solve, first collect like terms to obtain $2x + 1 = 437$. Then, subtract 1 from both sides and then divide by 2. The solution to the equation is $x = 218$. Therefore, the two consecutive page numbers that satisfy the problem are 218 and 219. It is always important to make sure that answers to real-world problems make sense. For instance, if the solution to this same problem resulted in decimals, that should be a red flag indicating the need to check the work. Page numbers are whole numbers; therefore, if decimals are found to be answers, the solution process should be double-checked to see where mistakes were made.

on is the process of breaking up a mathematical quantity, such as a number or polynomial, ͘.ouuct of two or more factors. For example, a factorization of the number 16 is $16 = 8 \times 2$. If multiplied out, the factorization results in the original number. A **prime factorization** is a specific factorization when the number is factored completely using prime numbers only. For example, the prime factorization of 16 is:

$$16 = 2 \times 2 \times 2 \times 2$$

A factor tree can be used to find the prime factorization of any number.

Within a factor tree, pairs of factors are found until no other factors can be used, as in the following factor tree of number 84:

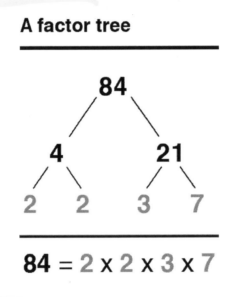

A factor tree

$$84 = 2 \times 2 \times 3 \times 7$$

It first breaks 84 into 21×4, which is not a prime factorization. Then, both 21 and 4 are factored into their primes. The final numbers on each branch consist of the numbers within the prime factorization. Therefore:

$$84 = 2 \times 2 \times 3 \times 7$$

Factorization can be helpful in finding greatest common divisors and least common denominators.

Also, a factorization of an algebraic expression can be found. Throughout the process, a more complicated expression can be decomposed into products of simpler expressions. To factor a polynomial, first determine if there is a greatest common factor. If there is, factor it out. For example, $2x^2 + 8x$ has a greatest common factor of $2x$, so it can be written as $2x(x + 4)$. Once the greatest common monomial factor is factored out, if applicable, count the number of terms in the polynomial. If there are two terms, is it a difference of squares, a sum of cubes, or a difference of cubes?

If so, the following rules can be used:

$$a^2 - b^2 = (a + b)(a - b)$$

$$a^3 + b^3 = (a + b)(a^2 - ab + b^2)$$

$$a^3 - b^3 = (a + b)(a^2 + ab + b^2)$$

If there are three terms, and if the trinomial is a perfect square trinomial, it can be factored into the following:

$$a^2 + 2ab + b^2 = (a + b)^2$$

$$a^2 - 2ab + b^2 = (a - b)^2$$

If not, try factoring into a product of two binomials by trial and error into a form of:

$$(x + p)(x + q)$$

For example, to factor $x^2 + 6x + 8$, determine what two numbers have a product of 8 and a sum of 6. Those numbers are 4 and 2, so the trinomial factors into:

$$(x + 2)(x + 4)$$

Finally, if there are four terms, try factoring by grouping. First, group terms together that have a common monomial factor. Then, factor out the common monomial factor from the first two terms. Next, look to see if a common factor can be factored out of the second set of two terms that results in a common binomial factor. Finally, factor out the common binomial factor of each expression, for example:

$$xy - x + 5y - 5$$

$$x(y - 1) + 5(y - 1)$$

$$(y - 1)(x + 5)$$

After the expression is completely factored, check to see if the factorization is correct by multiplying to try to obtain the original expression. Factorizations are helpful in solving equations that consist of a polynomial set equal to 0. If the product of two algebraic expressions equals 0, then at least one of the factors is equal to 0. Therefore, factor the polynomial within the equation, set each factor equal to 0, and solve. For example, $x^2 + 7x - 18 = 0$ can be solved by factoring into:

$$(x + 9)(x - 2) = 0$$

Set each factor equal to 0, and solve to obtain $x = -9$ and $x = 2$.

Identifying Integers
Integers include zero, and both positive and negative numbers with no fractional component. Examples of integers are −3, 5, 120, −47, and 0. Numbers that are not integers include 1.3333, ½, −5.7, and 4 ½. Integers can be used to describe different real-world situations. If a scuba diver were to dive 50 feet down into the ocean, his position can be described as −50, in relation to sea level. If, while traveling in Denver, Colorado, a car has an elevation reading of 2300 feet, the integer 2300 can be used to describe the feet above sea level. Integers can be used in many different ways to describe situations with whole numbers and zero.

Integers can also be added and subtracted as situations change. If the temperature in the morning is 45 degrees, and it dropped to 33 degrees in the afternoon, the difference can be found by subtracting the integers 45 and 33 to get a change of 12 degrees. If a submarine was at a depth of 100 feet below sea level, then rose 35 feet, the new depth can be found by adding −100 to 35. The following equation can be used to model the situation with integers:

$$-100 + 35 \ -65$$

The answer of −65 reveals the new depth of the submarine, 65 feet below sea level.

Recognizing Rational Exponents

A **rational number** is any number that can be written as a fraction of two integers. Examples of rational numbers include $\frac{1}{2}, \frac{5}{4}$, and 8. The number 8 is rational because it can be expressed as a fraction also, $\frac{8}{1} = 8$. **Rational exponents** represent one way to show how roots are used to express multiplication of any number by itself. For example, 3^2 has a base of 3 and rational exponent of 2, or $\frac{2}{1}$. It can be rewritten as the square root of 3 raised to the first power, or $\sqrt[2]{3^1}$. Any number with a rational exponent can be written this way. The **numerator,** or number on top of the fraction, becomes the root and the **denominator,** or bottom number of the fraction, becomes the whole number exponent. Another example is $4^{\frac{3}{2}}$. It can be rewritten as the square root of four to the third power, or $\sqrt[2]{4^3}$. This can be simplified by performing the operations 4 to the third power, $4^3 = 4 \times 4 \times 4 = 64$, and then taking the square root of 64, $\sqrt[2]{64}$, which yields an answer of 8. Another way of stating the answer would be 4 to power $\frac{3}{2}$ is 8, or 4 times itself $\frac{3}{2}$ times is 8.

Understanding Vectors

A **vector** is something that has both magnitude and direction. A vector may sometimes be represented by a ray that has a length, for its magnitude, and a direction. As the magnitude of the vector increases, the length of the ray changes. The direction of the ray refers to the way that the magnitude is applied. The following image shows the placement and parts of a vector.

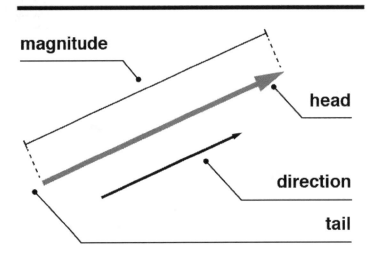

Parts of a Vector

magnitude

head

direction

tail

Examples of vector quantities include force and velocity. **Force** is a vector quantity because applying force requires magnitude, which is the amount of force, and a direction that the force is applied. **Velocity** is a vector because it has a magnitude, or speed that an object travels, and also the direction that the object is traveling in. Vectors can be added together by placing the tail of the second at the head of the first. The resulting vector is found by starting at the first tail and ending at the second head. An example of this is show in the following picture.

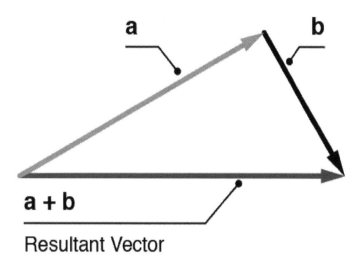

Subtraction can also be done with vectors by adding the inverse of the second vector. The inverse is found by reversing the direction of the vector. Then addition can take place just as described above, but using the inverse instead of the original vector. Scalar multiplication can also be done with vectors. This multiplication changes the magnitude of the vector by the **scalar,** or number. For example, if the length is described as 4, then scalar multiplication is used to multiply by 2, so the vector magnitude becomes 8. The direction of the vector is left unchanged because scalar does not include direction.

Vectors may also be described using coordinates on a plane, such as (5, 2). This vector would start at a point and move to the right 5 and up 2. The two coordinates describe the horizontal and vertical components of the vector. The starting point in relation to the coordinates is the tail, and the ending point is the head.

Creating Matrices
A **matrix** is an arrangement of numbers in rows and columns. Matrices are used to work with vectors and transform them. One example is a system of linear equations. Matrices can represent a system and be used to transform and solve the system. An important connection between scalars, vectors, and matrices is this: scalars are only numbers, vectors are numbers with magnitude and direction, and matrices are an array of numbers in rows and columns. The rows run from left to right and the columns run from top to bottom. When describing the dimensions of a matrix, the number of rows is stated first and the number of columns is stated second. The following matrix has two rows and three columns, referred to as a 2×3 matrix: $\begin{smallmatrix} 3 & 5 & 7 \\ 4 & 2 & 8 \end{smallmatrix}$. A number in a matrix can be found by describing its location. For example, the number in row two, column three is 8. In row one, column two, the number 5 is found.

Operations can be performed on matrices, just as they can on vectors. Scalar multiplication can be performed on matrices and it will change the magnitude, just as with a vector. A scalar multiplication problem using a 2×2 matrix looks like the following:

$$3 \times \begin{bmatrix} 4 & 5 \\ 8 & 3 \end{bmatrix}$$

The scalar of 3 is multiplied by each number to form the resulting matrix:

$$\begin{bmatrix} 12 & 15 \\ 24 & 9 \end{bmatrix}$$

Matrices can also be added and subtracted. For these operations to be performed, the matrices must be the same dimensions. Other operations that can be performed to manipulate matrices are multiplication, division, and transposition. **Transposing** a matrix means to switch the rows and columns. If the original matrix has two rows and three columns, then the transposed matrix has three rows and two columns.

Algebra

Solving, Graphing, and Modeling Multiple Types of Expressions

When presented with a real-world problem that must be solved, the first step is always to determine what the unknown quantity is that must be solved for. Use a **variable**, such as x or t, to represent that unknown quantity. Sometimes there can be two or more unknown quantities. In this case, either choose an additional variable, or if a relationship exists between the unknown quantities, express the other quantities in terms of the original variable. After choosing the variables, form algebraic expressions and/or equations that represent the verbal statement in the problem. The following table shows examples of vocabulary used to represent the different operations:

Addition	Sum, plus, total, increase, more than, combined, in all
Subtraction	Difference, less than, subtract, reduce, decrease, fewer, remain
Multiplication	Product, multiply, times, part of, twice, triple
Division	Quotient, divide, split, each, equal parts, per, average, shared

The combination of operations and variables form both mathematical expression and equations. The difference between expressions and equations are that there is no equals sign in an expression, and that expressions are **evaluated** to find an unknown quantity, while equations are **solved** to find an unknown quantity. Also, inequalities can exist within verbal mathematical statements. Instead of a statement of equality, expressions state quantities are *less than, less than or equal to, greater than,* or *greater than or equal to.* Another type of inequality is when a quantity is said to be *not equal to* another quantity. The symbol used to represent "not equal to" is ≠.

The steps for solving inequalities in one variable are the same steps for solving equations in one variable. The addition and multiplication principles are used. However, to maintain a true statement when using the $<, \leq, >$, and $\geq$ symbols, if a negative number is either multiplied times both sides of an inequality or divided from both sides of an inequality, the sign must be flipped. For instance, consider the following inequality: $3 - 5x \leq 8$. First, 3 is subtracted from each side to obtain $-5x \leq 5$. Then, both sides are divided by -5, while flipping the sign, to obtain $x \geq -1$. Therefore, any real number greater than or equal to -1 satisfies the original inequality.

Adding and Subtracting Linear Algebraic Expressions

To add and subtract linear algebra expressions, you must combine like terms. **Like terms** are described as those terms that have the same variable with the same exponent. In the following example, the x-terms can be added because the variable is the same and the exponent on the variable of one is also the same. These terms add to be $9x$. The other like terms are called **constants** because they have no variable component. These terms will add to be nine.

Example: Add $(3x - 5) + (6x + 14)$

$3x - 5 + 6x + 14$ Rewrite without parentheses

$3x + 6x - 5 + 14$ Commutative property of addition

$9x + 9$ Combine like terms

When subtracting linear expressions, be careful to add the opposite when combining like terms. Do this by distributing -1, which is multiplying each term inside the second parenthesis by negative one. Remember that distributing -1 changes the sign of each term.

Example: Subtract $(17x + 3) - (27x - 8)$

$17x + 3 - 27x + 8$ Distributive Property

$17x - 27x + 3 + 8$ Commutative property of addition

$-10x + 11$ Combine like terms

Example: Simplify by adding or subtracting:

$(6m + 28z - 9) + (14m + 13) - (-4z + 8m + 12)$

$6m + 28z - 9 + 14m + 13 + 4z - 8m - 12$ Distributive Property

$6m + 14m - 8m + 28z + 4z - 9 + 13 - 12$ Commutative Property of Addition

$12m + 32z - 8$ Combine like terms

Solving Problems Using Numerical and Algebraic Expressions

Translating sentences describing relationships between variables and constants to algebraic expressions and equations involves recognizing key words that represent mathematical operations. This process is known as **modeling.** For simplicity, let x be the variable, or the unknown quantity. Statements that include the four operations addition, subtraction, multiplication, and division exist in sentences that model linear relationships. For example, words and phrases that represent addition are "sum," "more than," and "increased by." Words and phrases that represent subtraction are "minus," "decreased by," "subtracted from," "difference," "less", "fewer than," and "less than." Words and phrases that represent multiplication are "times," "product of," "twice," "double," and "triple." Finally, words and phrases that represent division are "divided by," "quotient," and "reciprocal."

Most of the time, these words and phrases are combined to represent expressions that deal with one or more operation. For example, "ten subtracted from nine times a number" would be represented as $9x - 10$, and "the quotient of a number and 7 increased by 8" would be represented as $\frac{x}{7} + 8$. The word problems that typically use these expressions will have a statement of equality. For instance, the problem could say "ten subtracted from nine times a number equals 20; find that number." In this case, the algebraic expression shown previously would be set equal to 20 and then solved for x.

$$9x - 10 = 20$$

Add ten to both sides and then divide by 9 to get the solution $x = \frac{30}{9}$, which reduces to $x = \frac{10}{3}$.

Other types of expressions, besides linear expressions, can be the results of modeling. If the variable is raised to a power other than 1, the result is a **polynomial expression**. The path of an object thrown up into the air is a common example of this. The graph of an object represents an upside-down parabola, which is modeled by an equation of the type $y = -ax^2$. In this case, a represents the height of the object at its highest point before coming back down to the ground, and the negative sign shows that the parabola is upside down.

Here is the graph of a parabola:

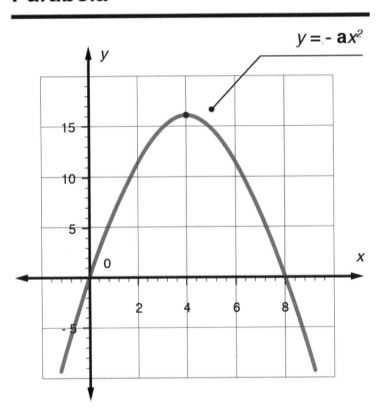

Evaluating and Simplifying Algebraic Expressions
Given an algebraic expression, students may be asked to evaluate for given values of variable(s). In doing so, students will arrive at a numerical value as an answer. For example:

Evaluate $a - 2b + ab$ for $a = 3$ and $b = -1$

To evaluate an expression, the given values should be substituted for the variables and simplified using the order of operations. In this case:

$$(3) - 2(-1) + (3)(-1)$$

Parentheses are used when substituting.

Given an algebraic expression, students may be asked to simplify the expression. For example:

Simplify:

$$5x^2 - 10x + 2 - 8x^2 + x - 1$$

Simplifying algebraic expressions requires combining like terms. A term is a number, variable, or product of a number and variables separated by addition and subtraction. The terms in the above expressions are: $5x^2, -10x, 2, -8x^2, x$, and -1. Like terms have the same variables raised to the same powers (exponents). To combine like terms, the coefficients (numerical factor of the term including sign) are added, while the variables and their powers are kept the same. The example above simplifies to:

$$-3x^2 - 9x + 1$$

Generating Equivalent Expressions
Two algebraic expressions are equivalent if, even though they look different, they represent the same expression. Therefore, plugging in the same values into the variables in each expression will result in the same result in both expressions. To obtain an equivalent form of an algebraic expression, laws of algebra must be followed. For instance, addition and multiplication are both commutative and associative. Therefore, terms in an algebraic expression can be added in any order and multiplied in any order. For instance, $4x + 2y$ is equivalent to $2y + 4x$ and:

$$y \times 2 + x \times 4$$

Also, the distributive law allows a number to be distributed throughout parentheses, as in the following: $a(b + c) = ab + ac$. The two expressions on both sides of the equals sign are equivalent. Also, collecting like terms is important when working with equivalent forms. The simplest version of an expression is always the one easiest to work with, so all like terms (those with the same variables raised to the same powers) must be combined.

Note that an expression is not an equation; therefore, expressions cannot be multiplied by numbers, divided by numbers, or have numbers added to them or subtracted from them and still have equivalent expressions. These processes can only happen in equations when the same step is performed on both sides of the equals sign.

The **distributive property** ($a(b + c) = ab + ac$) is a way of taking a factor and multiplying it through a given expression in parentheses. Each term inside the parentheses is multiplied by the outside factor, eliminating the parentheses. The following example shows how to distribute the number 3 to all the terms inside the parentheses.

Example: Use the distributive property to write an equivalent algebraic expression:

$3(2x + 7y + 6)$

$3(2x) + 3(7y) + 3(6)$ Distributive property

$6x + 21y + 18$ Simplify

Because $a - b$ can be written $a + (-b)$, the distributive property can be applied in the example below:

Example: Use the distributive property to write an equivalent algebraic expression.

$7(5m - 8)$

$7[5m + (-8)]$ Rewrite subtraction as addition of -8

$7(5m) + 7(-8)$ Distributive property

$35m - 56$ Simplify

In the following example, note that the factor of 2 is written to the right of the parentheses but is still distributed as before.

Example: Use the distributive property to write an equivalent algebraic expression:

$(3m + 4x - 10)2$

$(3m)2 + (4x)2 + (-10)2$ Distributive property

$6m + 8x - 20$ Simplify

Example: $-(-2m + 6x)$

In this example, the negative sign in front of the parentheses can be interpreted as $-1(-2m + 6x)$

$-1(-2m + 6x)$

$-1(-2m) + (-1)(6x)$ Distributive property

$2m - 6x$ Simplify

Linear Equations

An **equation in one variable** is a mathematical statement where two algebraic expressions with one variable, usually x, are set equal. To solve the equation, the variable must be isolated on one side of the equals sign. The addition and multiplication principles of equality are used to isolate the variable. The **addition principle of equality** states that the same number can be added to or subtracted from both sides of an equation. Because the same value is being used on both sides of the equals sign, equality is maintained. For example, the equation $2x = 5x$ is equivalent to both $2x + 3 = 5x + 3$, and $2x - 5 = 5x - 5$. This principle can be used to solve the following equation:

$$x + 5 = 4$$

The variable x must be isolated, so to move the 5 from the left side, subtract 5 from both sides of the equals sign. Therefore:

$$x + 5 - 5 = 4 - 5$$

So, the solution is $x = -1$. This process illustrates the idea of an **additive inverse** because subtracting 5 is the same as adding -5. Basically, add the opposite of the number that must be removed to both sides of the equals sign. The **multiplication principle of equality** states that equality is maintained when a number is either multiplied times both expressions on each side of the equals sign, or when both expressions are divided by the same number. For example, $4x = 5$ is equivalent to both $16x = 20$ and $x = \frac{5}{4}$. Multiplying both sides by 4 and dividing both sides by 4 maintains equality. Solving the equation $6x - 18 = 5$ requires the use of both principles. First, apply the addition principle to add 18 to both sides of the equals sign, which results in $6x = 23$. Then use the multiplication principle to divide both sides by 6, giving the solution $x = \frac{23}{6}$. Using the multiplication principle in the solving process is the same as involving a multiplicative inverse. A **multiplicative inverse** is a value that, when multiplied by a given number, results in 1. Dividing by 6 is the same as multiplying by $\frac{1}{6}$, which is both the reciprocal and multiplicative inverse of 6.

When solving a linear equation in one variable, checking the answer shows if the solution process was performed correctly. Plug the solution into the variable in the original equation. If the result is a false statement, something was done incorrectly during the solution procedure. Checking the example above gives the following:

$$6 \times \frac{23}{6} - 18$$

$$23 - 18 = 5$$

Therefore, the solution is correct.

Some equations in one variable involve fractions or the use of the distributive property. In either case, the goal is to obtain only one variable term and then use the addition and multiplication principles to isolate that variable. Consider the equation:

$$\frac{2}{3}x = 6$$

To solve for x, multiply each side of the equation by the reciprocal of $\frac{2}{3}$, which is $\frac{3}{2}$. This step results in $\frac{3}{2} \times \frac{2}{3}x = \frac{3}{2} \times 6$, which simplifies into the solution $x = 9$. Now consider the equation:

$$3(x + 2) - 5x = 4x + 1$$

Use the distributive property to clear the parentheses. Therefore, multiply each term inside the parentheses by 3. This step results in:

$$3x + 6 - 5x = 4x + 1$$

Next, collect like terms on the left-hand side. **Like terms** are terms with the same variable or variables raised to the same exponent(s). Only like terms can be combined through addition or subtraction. After collecting like terms, the equation is:

$$-2x + 6 = 4x + 1$$

29

apply the addition and multiplication principles. Add $2x$ to both sides to obtain:

$$6 = 6x + 1$$

Then, subtract 1 from both sides to obtain $5 = 6x$. Finally, divide both sides by 6 to obtain the solution"

$$\frac{5}{6} = x$$

Two other types of solutions can be obtained when solving an equation in one variable. The final result could be that there is either no solution or that the solution set contains all real numbers. Consider the equation:

$$4x = 6x + 5 - 2x$$

First, the like terms can be combined on the right to obtain:

$$4x = 4x + 5$$

Next, subtract $4x$ from both sides. This step results in the false statement $0 = 5$. There is no value that can be plugged into x that will ever make this equation true. Therefore, there is no solution. The solution procedure contained correct steps, but the result of a false statement means that no value satisfies the equation. The symbolic way to denote that no solution exists is Ø. Next, consider the equation:

$$5x + 4 + 2x = 9 + 7x - 5$$

Combining the like terms on both sides results in:

$$7x + 4 = 7x + 4$$

The left-hand side is exactly the same as the right-hand side. Using the addition principle to move terms, the result is $0 = 0$, which is always true. Therefore, the original equation is true for any number, and the solution set is all real numbers. The symbolic way to denote such a solution set is $\mathbb{R}$, or in interval notation, $(-\infty, \infty)$.

Algebraically Solving Linear Equations or Inequalities in One Variable
A *linear equation in one variable* can be solved using the following steps:

1. Simplify the algebraic expressions on both sides of the equals sign by removing all parentheses, using the distributive property, and then collect all like terms.

2. Collect all variable terms on one side of the equals sign and all constant terms on the other side by adding the same quantity to both sides of the equals sign, or by subtracting the same quantity from both sides of the equals sign.

3. Isolate the variable by either dividing both sides of the equation by the same number, or by multiplying both sides by the same number.

4. Check the answer.

The only difference between solving linear inequalities versus equations is that when multiplying by a negative number or dividing by a negative number, the direction of the inequality symbol must be reversed.

If an equation contains multiple fractions, it might make sense to clear the equation of fractions first by multiplying all terms by the least common denominator. Also, if an equation contains several decimals, it might make sense to clear the decimals as well by multiplying times a factor of 10. If the equation has decimals in the hundredth place, multiply every term in the equation by 100.

Polynomial Equations

A **polynomial** is a mathematical expression containing the sum and difference of one or more terms that are constants multiplied times variables raised to positive powers. A polynomial is considered expanded when there are no variables contained within parentheses, the distributive property has been carried out for any terms that were within parentheses, and like terms have been collected.

When working with polynomials, **like terms** are terms that contain exactly the same variables with the same powers. For example, x^4y^5 and $9x^4y^5$ are like terms. The coefficients are different, but the same variables are raised to the same powers. When adding polynomials, only terms that are considered like terms can be added. When adding two like terms, just add the coefficients and leave the variables alone. This process uses the distributive property. For example:

$$x^4y^5 + 9x^4y^5$$
$$(1+9)x^4y^5$$
$$10x^4y^5$$

Therefore, when adding two polynomials, simply add the like terms together. Unlike terms cannot be combined.

Subtracting polynomials involves adding the opposite of the polynomial being subtracted. Basically, the sign of each term in the polynomial being subtracted is changed, and then the like terms are combined because it is now an addition problem. For example, consider the following:

$$6x^2 - 4x + 2 - (4x^2 - 8x + 1)$$

Add the opposite of the second polynomial to obtain:

$$6x^2 - 4x + 2 + (-4x^2 + 8x - 1)$$

Then, collect like terms to obtain:

$$2x^2 + 4x + 1$$

Multiplying polynomials involves using the product rule for exponents that:

$$b^m b^n = b^{m+n}$$

Basically, when multiplying expressions with the same base, just add the exponents. Multiplying a monomial by a monomial involves multiplying the coefficients together and then multiplying the variables together using the product rule for exponents. For instance:

$$8x^2y \times 4x^4y^2 = 32x^6y^3$$

When multiplying a monomial by a polynomial that is not a monomial, use the distributive property to multiply each term of the polynomial times the monomial. For example:

$$3x(x^2 + 3x - 4) = 3x^3 + 9x^2 - 12x$$

31

Finally, multiplying two polynomials when neither one is a monomial involves multiplying each term of the first polynomial by each term of the second polynomial. There are some shortcuts, given certain scenarios. For instance, a binomial times a binomial can be found by using the **FOIL (Firsts, Outers, Inners, Lasts)** method shown here.

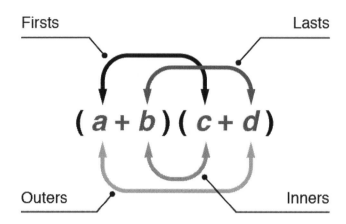

Finding the product of a sum and difference of the same two terms is simple because if it was to be foiled out, the outer and inner terms would cancel out. For instance:

$$(x + y)(x - y) = x^2 + xy - xy - y^2$$

Finally, the square of a binomial can be found using the following formula:

$$(a \pm b)^2 = a^2 \pm 2ab + b^2$$

Radical and Exponential Relationships

The nth root of a is given as $\sqrt[n]{a}$, which is called a **radical.** Typical values for n are 2 and 3, which represent the square and cube roots. In this form, n represents an integer greater than or equal to 2, and a is a real number. If n is even, a must be nonnegative, and if n is odd, a can be any real number. This radical can be written in exponential form as $a^{\frac{1}{n}}$. Therefore, $\sqrt[4]{15}$ is the same as $15^{\frac{1}{4}}$ and $\sqrt[3]{-5}$ is the same as $(-5)^{\frac{1}{3}}$.

In a similar fashion, the nth root of a can be raised to a power m, which is written as:

$$\left(\sqrt[n]{a}\right)^m$$

This expression is the same as:

$$\sqrt[n]{a^m}$$

For example:

$$\sqrt[2]{4^3} = \sqrt[2]{64} = 8 = \left(\sqrt[2]{4}\right)^3 = 2^3$$

Because $\sqrt[n]{a} = a^{\frac{1}{n}}$, both sides can be raised to an exponent of m, resulting in:

$$\left(\sqrt[n]{a}\right)^m = \sqrt[n]{a^m} = a^{\frac{m}{n}}$$

This rule allows:

$$\sqrt[2]{4^3} = \left(\sqrt[2]{4}\right)^3$$

$$4^{\frac{3}{2}} = (2^2)^{\frac{3}{2}}$$

$$2^{\frac{6}{2}} = 2^3 = 8$$

Negative exponents can also be incorporated into these rules. Any time an exponent is negative, the base expression must be flipped to the other side of the fraction bar and rewritten with a positive exponent. For instance:

$$2^{-3} = \frac{1}{2^3} = \frac{1}{8}$$

Therefore, two more relationships between radical and exponential expressions are:

$$a^{-\frac{1}{n}} = \frac{1}{\sqrt[n]{a}}$$

and:

$$a^{-\frac{m}{n}} = \frac{1}{\sqrt[n]{a^m}} = \frac{1}{\left(\sqrt[n]{a}\right)^m}$$

Thus:

$$8^{-3} = \frac{1}{\sqrt[3]{8}} = \frac{1}{2}$$

All of these relationships are very useful when simplifying complicated radical and exponential expressions. If an expression contains both forms, use one of these rules to change the expression to contain either all radicals or all exponential expressions. This process makes the entire expression much easier to work with, especially if the expressions are contained within equations.

Consider the following example:

$$\sqrt{x} \times \sqrt[4]{x}$$

It is written in radical form; however, it can be simplified into one radical by using exponential expressions first. The expression can be written as:

$$x^{\frac{1}{2}} \times x^{\frac{1}{4}}$$

It can be combined into one base by adding the exponents as:

$$x^{\frac{1}{2}+\frac{1}{4}} = x^{\frac{3}{4}}$$

Writing this back in radical form, the result is:

$$\sqrt[4]{x^3}$$

Creating, Solving, or Interpreting Systems of Linear Inequalities in Two Variables

A **system of linear inequalities in two variables** consists of two inequalities in two variables, typically *x* and *y*. For example, the following is a system of linear inequalities in two variables:

$$\begin{cases} 4x + 2y < 1 \\ 2x - y \leq 0 \end{cases}$$

The curly brace on the left side shows that the two inequalities are grouped together. A solution of a single inequality in two variables is an ordered pair that satisfies the inequality. For example, (1, 3) is a solution of the linear inequality $y \geq x + 1$ because when plugged in, it results in a true statement. The graph of an inequality in two variables consists of all ordered pairs that make the solution true. Therefore, the entire solution set of a single inequality contains many ordered pairs, and the set can be graphed by using a half plane.

A **half plane** consists of the set of all points on one side of a line. If the inequality consists of > or <, the line is dashed because no solutions actually exist on the line shown. If the inequality consists of ≥ or ≤, the line is solid and solutions are on the line shown. To graph a linear inequality, graph the corresponding equation found by replacing the inequality symbol with an equals sign. Then pick a test point that exists on either side of the line. If that point results in a true statement when plugged into the original inequality, shade in the side containing the test point. If it results in a false statement, shade in the opposite side.

Solving a system of linear inequalities must be done graphically. Follow the process as described above for both given inequalities. The solution set to the entire system is the region that is in common to every graph in the system. For example, here is the solution to the following system:

$$\begin{cases} y \geq 3 - x \\ y \leq -3 - x \end{cases}$$

The solution to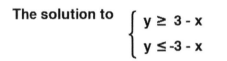

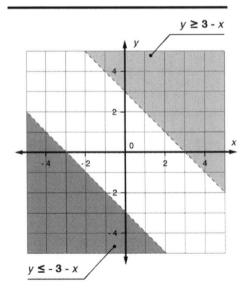

34

Note that there is no region in common, so this system has no solution.

Creating, Solving, or Interpreting Systems of Two Linear Equations in Two Variables
An example of a system of two linear equations in two variables is the following:

$$2x + 5y = 8$$

$$5x + 48y = 9$$

A solution to a **system of two linear equations** is an ordered pair that satisfies both the equations in the system. A system can have one solution, no solution, or infinitely many solutions. The solution can be found through a graphing technique. The solution of a system of equations is actually equal to the point of intersection of both lines. If the lines intersect at one point, there is one solution and the system is said to be **consistent**. However, if the two lines are parallel, they will never intersect and there is no solution. In this case, the system is said to be **inconsistent**. Thirdly, if the two lines are actually the same line, there are infinitely many solutions and the solution set is equal to the entire line. The lines are **dependent**.

Here is a summary of the three cases:

Solving Systems by Graphing

Consistent	**Inconsistent**	**Dependent**
One solution	No solution	Infinite number of solutions
Lines intersect	*Lines are parallel*	*Coincide: same line*

Consider the following system of equations:

$$\begin{cases} y + x = 3 \\ y - x = 1 \end{cases}$$

To find the solution graphically, graph both lines on the same *xy*-plane. Graph each line using either a table of ordered pairs, the *x*- and *y*-intercepts, or slope and the *y*-intercept. Then, locate the point of intersection.

The graph is shown here:

The System of Equations $\begin{cases} y + x = 3 \\ y - x = 1 \end{cases}$

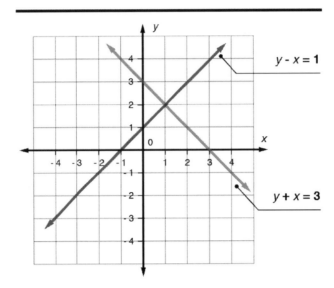

It can be seen that the point of intersection is the ordered pair (1, 2). This solution can be checked by plugging it back into both original equations to make sure it results in true statements. This process results in:

$$2 + 1 = 3$$

$$2 - 1 = 1$$

Both are true equations, so therefore the point of intersection is truly the solution.

The following system has no solution:

$$y = 4x + 1$$

$$y = 4x - 1$$

Both lines have the same slope and different y-intercepts, and therefore they are parallel. This means that they run alongside each other and never intersect.

Finally, the following solution has infinitely many solutions:

$$2x - 7y = 12$$

$$4x - 14y = 24$$

Note that the second equation is equal to the first equation times 2. Therefore, they are the same line. The solution set can be written in set notation as $\{(x, y) | 2x - 7y = 12\}$, which represents the entire line.

Algebraically Solving Systems of Two Linear Equations in Two Variables

There are two algebraic methods to finding solutions. The first is **substitution**. This process is
suited for systems when one of the equations is already solved for one variable, or when solving
variable is easy to do. The equation that is already solved for is substituted into the other equation
that variable, and this process results in a linear equation in one variable. This equation can be sol
the given variable, and then that solution can be plugged into one of the original equations, which
then be solved for the other variable. This last step is known as **back-substitution** and the end result is an
ordered pair.

A system that is best suited for substitution is the following:

$$y = 4x + 2$$

$$2x + 3y = 9$$

The other method is known as **elimination,** or the **addition method**. This is better suited when the
equations are in standard form:

$$Ax + By = C$$

The goal in this method is to multiply one or both equations times numbers that result in opposite
coefficients. Then, add the equations together to obtain an equation in one variable. Solve for the given
variable, then take that value and back-substitute to obtain the other part of the ordered pair solution.

A system that is best suited for elimination is the following:

$$2x + 3y = 8$$

$$4x - 2y = 10$$

Note that in order to check an answer when solving a system of equations, the solution must be checked
in both original equations to show that it solves not only one of the equations, but both of them.

If throughout either solution procedure the process results in an untrue statement, there is no solution to
the system. Finally, if throughout either solution procedure the process results in the variables dropping
out, which gives a statement that is always true, there are infinitely many solutions.

Functions

Function Definition

A **relation** is any set of ordered pairs (x, y). The first set of points, known as the *x*-coordinates, make up
the **domain** of the relation. The second set of points, known as the *y*-coordinates, make up the **range** of
the relation. A relation in which every member of the domain corresponds to only one member of the
range is known as a **function**. A function cannot have a member of the domain corresponding to two
members of the range.

Function Notation

Functions are most often given in terms of equations instead of ordered pairs. For instance, here is an
equation of a line:

$$y = 2x + 4$$

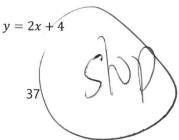

37

In function notation, this can be written as:

$$f(x) = 2x + 4$$

The expression $f(x)$ is read "f of x" and it shows that the inputs, the x-values, get plugged into the function and the output is:

$$y = f(x)$$

The set of all inputs are in the domain and the set of all outputs are in the range.

The x-values are known as the **independent variables** of the function and the y-values are known as the **dependent variables** of the function. The y-values depend on the x-values. For instance, if $x = 2$ is plugged into the function shown above, the y-value depends on that input.

$$f(2) = 2 \times 2 + 4 = 8$$

Therefore, $f(2) = 8$, which is the same as writing the ordered pair (2, 8). To graph a function, graph it in equation form. Therefore, replace $f(x)$ with h and plot ordered pairs.

Due to the definition of a function, the graph of a function cannot have two first components (x-values) being paired to the same second component. Therefore, all graphs of functions pass the **vertical line test.** If any vertical line intersects a graph in more than one place, the graph is not that of a function. For instance, the graph of a circle is not a function because one can draw a vertical line through a circle and the would intersect the circle twice. Common functions include lines that aren't vertical and polynomials, and they all pass the vertical line test.

Representing Functions
When presented with a real-world problem that must be solved, the first step is always to determine what the unknown quantity is that must be solved for. Use a variable, such as x or t, to represent that unknown quantity. Sometimes, there can be two or more unknown quantities. In this case, either choose an additional variable, or if a relationship exists between the unknown quantities, express the other quantities in terms of the original variable. After choosing the variables, form algebraic expressions and/or equations that represent the verbal statement in the problem. The following table shows examples of vocabulary used to represent the different operations:

Addition	Sum, plus, total, increase, more than, combined, in all
Subtraction	Difference, less than, subtract, reduce, decrease, fewer, remain
Multiplication	Product, multiply, times, part of, twice, triple
Division	Quotient, divide, split, each, equal parts, per, average, shared

The combination of operations and variables form both mathematical expression and equations. The difference between expressions and equations are that there is no equals sign in an expression, and that expressions are **evaluated** to find an unknown quantity, while equations are **solved** to find an unknown quantity. Also, inequalities can exist within verbal mathematical statements. Instead of a statement of equality, expressions state quantities are *less than, less than or equal to, greater than,* or *greater than or equal to.* Another type of inequality is when a quantity is said to be *not equal to* another quantity. The symbol used to represent "not equal to" is $\neq$.

The steps for solving inequalities in one variable are the same steps for solving equations in one variable. The addition and multiplication principles are used. However, to maintain a true statement when using the $<, \leq, >$, and $\geq$ symbols, if a negative number is either multiplied by both sides of an inequality or divided from both sides of an inequality, the sign must be flipped. For instance, consider the following inequality:

$$3 - 5x \leq 8$$

First, 3 is subtracted from each side to obtain:

$$-5x \leq 5$$

Then, both sides are divided by -5, while flipping the sign, to obtain:

$$x \geq -1$$

Therefore, any real number greater than or equal to -1 satisfies the original inequality.

Linear Functions
A linear function that models a linear relationship between two quantities is of the form $y = mx + b$, or in function form $f(x) = mx + b$. In a linear function, the value of y depends on the value of x, and y increases or decreases at a constant rate as x increases. Therefore, the independent variable is x, and the dependent variable is y. The graph of a linear function is a line, and the constant rate can be seen by looking at the steepness, or slope, of the line. If the line increases from left to right, the slope is positive. If the line slopes downward from left to right, the slope is negative.

In the function, m represents slope. Each point on the line is an **ordered pair** (x, y), where x represents the x-coordinate of the point and y represents the y-coordinate of the point. The point where $x = 0$ is known as the y-intercept, and it is the place where the line crosses the y-axis. If $x = 0$ is plugged into $f(x) = mx + b$, the result is $f(0) = b$; therefore, the point $(0, b)$ is the y-intercept of the line. The derivative of a linear function is its slope.

Consider the following situation. A taxicab driver charges a flat fee of $2 per ride and $3 a mile. This statement can be modeled by the function $f(x) = 3x + 2$ where x represents the number of miles and $f(x) = y$ represents the total cost of the ride. The total cost increases at a constant rate of $2 per mile, which is why this situation is a linear relationship. The slope $m = 3$ is equivalent to this rate of change. The flat fee of $2 is the y-intercept. It is the place where the graph crosses the x-axis, and it represents the cost when $x = 0$, or when no miles have been traveled in the cab. The y-intercept in this situation represents the flat fee.

A linear function of the form $f(x) = mx + b$ has two important quantities: m and b. The quantity m represents the slope of the line, and the quantity b represents the y-intercept of the line. When the function represents an actual real-life situation, or mathematical model, these two quantities are very meaningful. The slope, m, represents the rate of change, or the amount y increases or decreases given an increase in x. If m is positive, the rate of change is positive, and if m is negative, the rate of change is negative. The y-intercept, b, represents the amount of the quantity y when x is 0. In many applications, if the x-variable is never a negative quantity, the y-intercept represents the initial amount of the quantity y. Often the x-variable represents time, so it makes sense that the x-variable is never negative.

Consider the following example. These two equations represent the cost, C, of t-shirts, x, at two different printing companies:

$$C(x) = 7x$$

$$C(x) = 5x + 25$$

The first equation represents a scenario that shows the cost per t-shirt is $7. In this equation, x varies directly with y. There is no y-intercept, which means that there is no initial cost for using that printing company. The rate of change is 7, which is price per shirt. The second equation represents a scenario that has both an initial cost and a cost per t-shirt. The slope 5 shows that each shirt is $5. The y-intercept 25 shows that there is an initial cost of $25 when using that company. Therefore, it makes sense to use the first company at $7 a shirt when only purchasing a small number of t-shirts. However, large orders would be cheaper by going with the second company because eventually that initial cost will become negligible.

Radical Functions

Recall that a **radical expression** is an expression involving a square root, a cube root, or a higher order root such as fourth root, fifth root, etc. The expression underneath the radical is known as the **radicand** and the index is the number corresponding to the **root**. An index of 2 corresponds to a square root. A radical function is a function that involves a radical expression. For instance, $\sqrt{x+1}$ is a radical expression, $x + 1$ is the radicand, and the corresponding function is:

$$y = \sqrt{x + 1}$$

The function can also be written in function notation as:

$$f(x) = \sqrt{x + 1}$$

If the root is even, meaning a square root, fourth root, etc., the radicand must be positive. Therefore, in order to find the domain of a radical function with an even index, set the radicand greater than or equal to zero and find the set of numbers that satisfies that inequality. The domain of $f(x) = \sqrt{x+1}$ is all numbers greater than or equal to -1. The range of this function is all nonnegative real numbers because the square root, or any even root, can never output a negative number. The domain of an odd root is all real numbers because the radicand can be negative in an odd root.

Piecewise Functions

A **piecewise function** is basically a function that is defined in pieces. The graph of the function behaves differently over different intervals along the x-axis, or different intervals of its domain. Therefore, the function is defined using different mathematical expressions over these intervals. The function is not defined by only one equation. In a piecewise function, the function is actually defined by two or more equations, where each equation is used over a specific interval of the domain.

Here is a graph of a piecewise function:

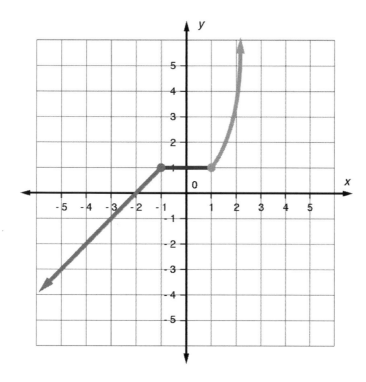

Notice that from $(-\infty, -1]$, the graph is a line with positive slope. From $[-1, 1]$ the graph is a horizontal line. Finally, from $[1, \infty)$ the graph is a nonlinear curve. Both the domain and range of this piecewise defined function is all real numbers, which is expressed as $(-\infty, \infty)$.

Piecewise functions can also have **discontinuities**, which are jumps in the graph. When drawing a graph, if the pencil must be picked up at any point to continue drawing, the graph has a discontinuity. Here is the graph of a piecewise function with discontinuities at $x = 1$ and $x = 2$:

A Piecewise Function

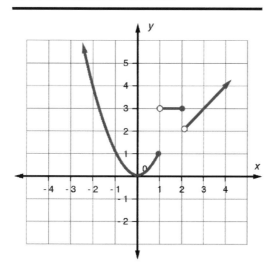

The open circle at a point indicates that the endpoint is not included in that part of the graph, and the closed circle indicates that the endpoint is included. The domain of this function is all real numbers; however, the range is all nonnegative real numbers $[0, \infty)$.

Polynomial Functions

A **polynomial function** is a function containing a polynomial expression, which is an expression containing constants and variables combined using the four mathematical operations. The degree of a polynomial in one variable is the largest exponent seen on any variable in the expression. Typical polynomial functions are **quartic,** with a degree of 4, **cubic,** with a degree of 3, and **quadratic,** with a degree of 2. Note that the exponents on the variables can only be nonnegative integers. The domain of any polynomial function is all real numbers because any number plugged into a polynomial expression grants a real number output. An example of a quartic polynomial equation is:

$$y = x^4 + 3x^3 - 2x + 1$$

The zeros of a polynomial function are the points where its graph crosses the y-axis. In order to find the number of real zeros of a polynomial function, **Descartes' Rule of Sign** can be used. The number of possible positive real zeros is equal to the number of sign changes in the coefficients of the terms in the polynomial. If there is only one sign change, there is only one positive real zero. In the example above, the signs of the coefficients are positive, positive, negative, and positive. Therefore, the sign changes two times and therefore, there are at most two positive real zeros. The number of possible negative real zeros is equal to the number of sign changes in the coefficients when plugging $-x$ into the equation. Again, if there is only one sign change, there is only one negative real zero. The polynomial result when plugging -x into the equation is:

$$y^4 - 3x^3 + 2x + 1$$

The sign changes two times, so there are at most two negative real zeros. Another polynomial equation this rule can be applied to is:

$$y = x^3 + 2x - x - 5$$

There is only one sign change in the terms of the polynomial, so there is exactly one real zero. When plugging -x into the equation, the polynomial result is:

$$-x^3 - 2x - x - 5$$

There are no sign changes in this polynomial, so there are no possible negative zeros.

Logarithmic Functions

For $x > 0, b > 0, b \neq 1$, the function $f(x) = \log_b x$ is known as the **logarithmic function** with base b. With $y = \log_b x$, its exponential equivalent is $b^y = x$. In either case, the exponent is y and the base is b. Therefore, $3 = \log_2 8$ is the same as $2^2 = 8$. So, in order to find the logarithm with base 2 of 8, find the exponent that when 2 is raised to that value results in 8. Similarly:

$$\log_3 243 = 5$$

In order to do this mentally, ask the question, what exponent does 3 need to be raised to that results in 243? The answer is 5. Most logarithms do not have whole number results. In this case, a calculator can be used. A calculator typically has buttons with base 10 and base e, so the change of base formula can be used to calculate these logs.

For instance:

$$\log_3 55 = \frac{\log 55}{\log 3} = 3.64$$

Similarly, the natural logarithm with base e could be used to obtain the same result.

$$\log_3 55 = \frac{\ln 55}{\ln 3} = 3.64$$

The domain of a logarithmic function $f(x) = \log_b x$ is all positive real numbers. This is because the exponent must be a positive number. The range of a logarithmic function $f(x) = \log_b x$ is all real numbers. The graphs of all logarithmic functions of the form $f(x) = \log_b x$ always pass through the point (1, 0) because anything raised to the power of 0 is 1. Therefore, such a function always has an *x*-intercept at 1. If the base is greater than 1, the graph increases from the left to the right along the *x*-axis. If the base is between 0 and 1, the graph decreases from the left to the right along the *x*-axis. In both situations, the *y*-axis is a vertical **asymptote**. The graph will never touch the *y*-axis, but it does approach it closely.

Here are the graphs of the two cases of logarithmic functions:

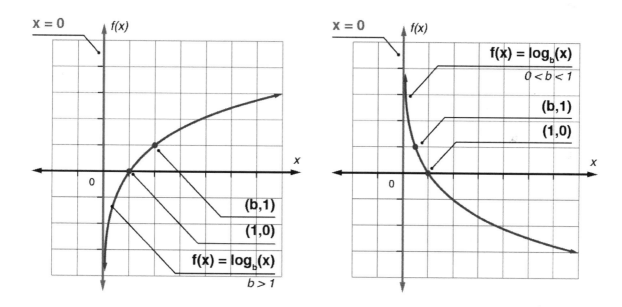

Finding and Applying Important Features of Graphs
A **graph** is a pictorial representation of the relationship between two variables. To read and interpret a graph, it is necessary to identify important features of the graph. First, read the title to determine what data sets are being related in the graph. Next, read the axis labels and understand the scale that is used. The horizontal axis often displays categories, like years, month, or types of pets. The vertical axis often displays numerical data like amount of income, number of items sold, or number of pets owned. Check to see what increments are used on each axis. The changes on the axis may represent fives, tens, hundreds, or any increment. Be sure to note what the increment is because it will affect the interpretation of the graph. Now, locate on the graph an element of interest and move across to find the element to which it relates. For example, notice an element displayed on the horizontal axis, find that element on the graph,

and then follow it across to the corresponding point on the vertical axis. Using the appropriate scale, interpret the relationship.

Choosing Appropriate Graphs to Display Data
Data may be displayed with a line graph, bar graph, or pie chart.

- A line graph is used to display data that changes continuously over time.

- A bar graph is used to compare data from different categories or groups and is helpful for recognizing relationships.

- A pie chart is used when the data represents parts of a whole.

<u>*Identifying, Constructing, and Completing Graphs That Correctly Represent Given Data*</u>
Data is often displayed with a line graph, bar graph, or pie chart.

The line graph below shows the number of push-ups that a student did over one week.

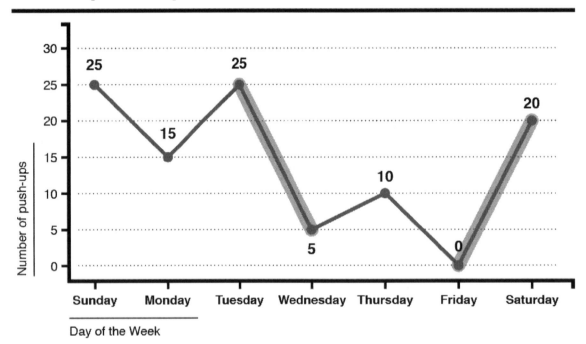

Notice that the horizontal axis displays the day of the week and the vertical axis displays the number of push-ups. A point is placed above each day of the week to show how many push-ups were done each day. For example, on Sunday the student did 25 push-ups. The line that connects the points shows how much the number of push-ups fluctuated throughout the week.

The bar graph below compares number of people who own various types of pets.

What kind of pet do you own?

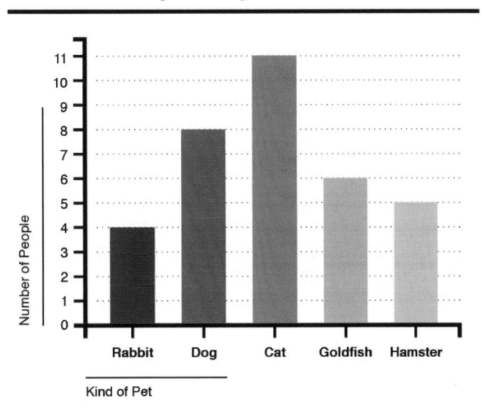

On the horizontal axis, the kind of pet is displayed. On the vertical axis, the number of people is displayed. Bars are drawn to show the number of people who own each type of pet. With the bar graph, it can quickly be determined that the fewest number of people own a rabbit and the greatest number of people own a cat.

The pie graph below displays students in a class who scored A, B, C, or D. Each slice of the pie is drawn to show the portion of the whole class that is represented by each letter grade. For example, the smallest portion represents students who scored a D. This means that the fewest number of students scored a D.

Student Grades

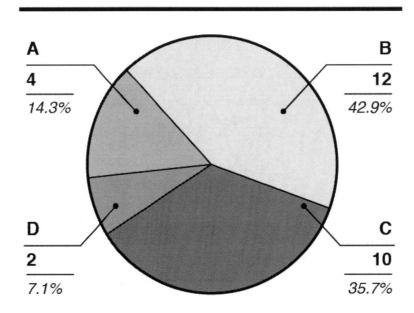

A
4
14.3%

B
12
42.9%

D
2
7.1%

C
10
35.7%

Geometry

Defining and Applying Knowledge of Shapes and Solids
Shapes are defined by their angles and number of sides. A shape with one continuous side, where all points on that side are equidistant from a center point is called a **circle.** A shape made with three straight line segments is a **triangle.** A shape with four sides is called a **quadrilateral,** but more specifically a **square, rectangle, parallelogram**, or **trapezoid,** depending on the interior angles. These shapes are two-dimensional and only made of straight lines and angles. **Solids** can be formed by combining these shapes and forming three-dimensional figures. These figures have another dimension because they add one more direction. Examples of solids include prisms or spheres.

There are four figures below that can be described based on their sides and dimensions. Figure 1 is a **cone** because it has three dimensions, where the bottom is a circle and the top is formed by the sides combining to one point. Figure 2 is a **triangle** because it has two dimensions, made up of three-line segments.

Figure 3 is a **cylinder** made up of two base circles and a rectangle to connect them in three dimensions. Figure 4 is an **oval** because it is one continuous line in two dimensions, not equidistant from the center.

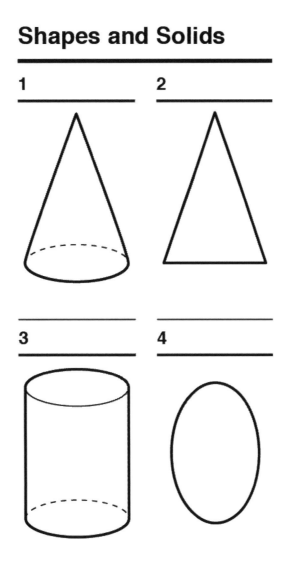

Shapes and Solids

1

2

3

4

Figure 5 below is made up of squares in three dimensions, combined to make a **cube**. Figure 6 is a rectangle because it has four sides that intersect at right angles. More specifically, it can be described as a **square** because the four sides have equal measures. Figure 7 is a **pyramid** because the bottom shape is a square and the sides are all triangles. These triangles intersect at a point above the square. Figure 8 is a **circle** because it is made up of one continuous line where the points are all equidistant from one center point.

Shapes and Solids

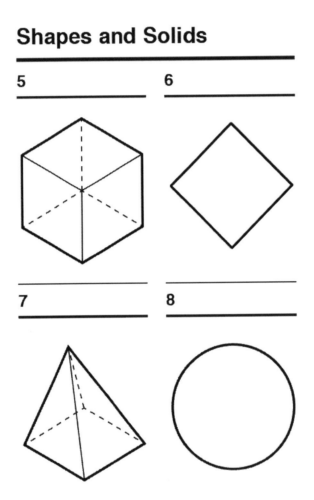

Congruence and Similarity

Two figures are **congruent** if they have the same shape and same size, meaning same angle measurements and equal side lengths. Two figures are **similar** if they have the same angle measurement but not side lengths. Basically, angles are congruent in similar triangles and their side lengths are constant multiplies of each other. Proving two shapes are similar involves showing that all angles are the same; proving two shapes are congruent involves showing that all angles are the same and that all sides are the same. If two pairs of angles are congruent in two triangles, then those triangles are similar because their third angle has to be equal due to the fact that all three angles add up to 180 degrees.

There are five main theorems that are used to show triangles are congruent. Each theorem involves showing different combinations of sides and angles are the same in two triangles, which proves the triangles are congruent. The **side-side-side (SSS) theorem** states that if all sides are equal in two triangles, the triangles are congruent. The **side-angle-side (SAS) theorem** states that if two pairs of sides are equal and the included angles are congruent in two triangles, then the triangles are congruent.

Similarly, the **angle-side-angle (ASA) theorem** states that if two pairs of angles are congruent and the included side lengths are equal in two triangles, the triangles are similar. The **angle-angle-side (AAS) theorem** states that two triangles are congruent if they have two pairs of congruent angles and a pair of corresponding equal side lengths that are not included. Finally, the **hypotenuse-leg (HL) theorem** states that if two right triangles have equal hypotenuses and an equal pair of shorter sides, the triangles are congruent. An important item to note is that angle-angle-angle (AAA) is not enough information to have congruence because if three angles are equal in two triangles, the triangles can only be described as similar.

Using the Relationship Between Similarity, Right Triangles, and Trigonometric Ratios
Within two similar triangles, corresponding side lengths are proportional, and angles are equal. In other words, regarding corresponding sides in two similar triangles, the ratio of side lengths is the same. Recall that the SAS theorem for similarity states that if an angle in one triangle is congruent to an angle in a second triangle, and the lengths of the sides in both triangles are proportional, then the triangles are similar. Also, because the ratio of two sides in two similar right triangles is the same, the trigonometric ratios in similar right triangles are always going to be equal.

If two triangles are similar, and one is a right triangle, the other is a right triangle. The definition of similarity ensures that each triangle has a 90-degree angle. In a similar sense, if two triangles are right triangles containing a pair of equal acute angles, the triangles are similar because the third pair of angles must be equal as well. However, right triangles are not necessarily always similar.

The following triangles are similar:

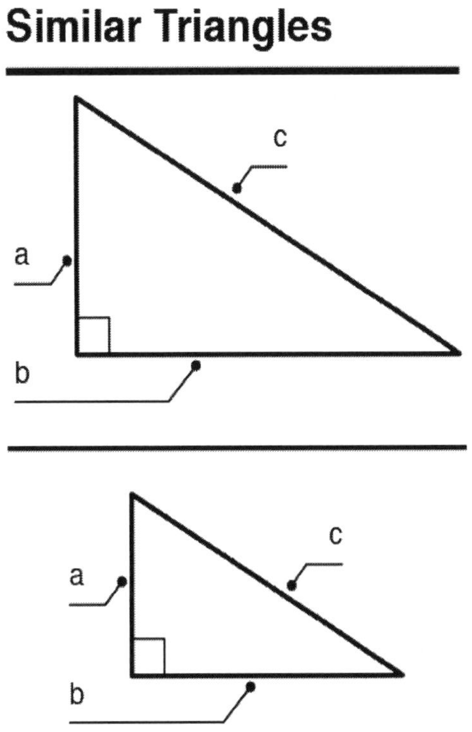

Similar Triangles

It is not always apparent at first glance, but theorems can be used to show similarity. The **Pythagorean Theorem** can be used to find the missing side lengths in both triangles. In the larger triangle, the missing side is the hypotenuse, c. Therefore:

$$9^2 + 12^2 = c^2$$

This equation is equivalent to $225 = c^2$, so taking the square root of both sides results in the positive root $c = 15$. In the other triangle, the Pythagorean Theorem can be used to find the missing side length b. The theorem shows that $6^2 + b^2 = 10^2$, and b is then solved for to obtain $b = 8$. The ratio of the sides in the larger triangle to the sides in the smaller triangle is the same value, 1.5. Therefore, the sides are proportional. Because they are both right triangles, they have a congruent angle. The SAS theorem for similarity can be used to show that these two triangles are similar.

Surface Area and Volume Measurements

Surface area is defined as the area of the surface of a figure. A **pyramid** has a surface made up of four triangles and one square. To calculate the surface area of a pyramid, the areas of each individual shape are calculated. Then the areas are added together. This method of decomposing the shape into two-dimensional figures to find area, then adding the areas, can be used to find surface area for any figure. Once these measurements are found, the area is described with square units. For example, the following figure shows a rectangular prism. The figure beside it shows the rectangular prism broken down into two-dimensional shapes, or rectangles. The area of each rectangle can be calculated by multiplying the length by the width. The area for the six rectangles can be represented by the following expression:

$$5 \times 6 + 5 \times 10 + 5 \times 6 + 6 \times 10 + 5 \times 10 + 6 \times 10$$

The total for all these areas added together is 280m^2, or 280 square meters.

This measurement represents the surface area because it is the area of all six surfaces of the rectangular prism.

The Net of a Rectangular Prism

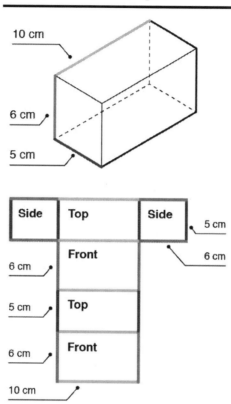

As mentioned, surface area of three-dimensional figures is the total area of each of the faces of the figures. Nets are used to lay out each face of an object. The following figure shows a triangular prism. The bases are triangles and the sides are rectangles. The second figure shows the net for this triangular prism. The dimensions are labeled for each of the faces of the prism. The area for each of the two triangles can be determined by the formula:

$$A = \frac{1}{2}bh$$

$$\frac{1}{2} \times 8 \times 9 = 36cm^2$$

The rectangle areas can be described by the equation:

$$A = lw = 8 \times 5 + 9 \times 5 + 10 \times 5$$

$$40 + 45 + 50 = 135cm^2$$

The area for the triangles can be multiplied by two, then added to the rectangle areas to yield a total surface are of 207cm².

A Triangular Prism and Its Net

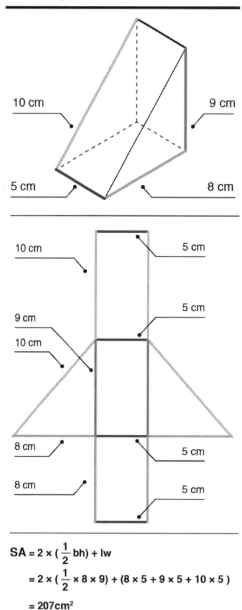

$$SA = 2 \times (\frac{1}{2} bh) + lw$$

$$= 2 \times (\frac{1}{2} \times 8 \times 9) + (8 \times 5 + 9 \times 5 + 10 \times 5)$$

$$= 207cm^2$$

Another shape that has a surface area is a cylinder. The shapes used to make up the **cylinder** are two circles and a rectangle wrapped around between the two circles. A common example of a cylinder is a can. The two circles that make up the bases are obvious shapes. The rectangle can be more difficult to see, but the label on a can will help illustrate it. When the label is removed from a can and laid flat, the shape is a rectangle.

When the areas for each shape are needed, there will be two formulas. The first is the area for the circles on the bases. This area is given by the formula $A = \pi r^2$. There will be two of these areas. Then the area of the rectangle must be determined. The width of the rectangle is equal to the height of the can, *h*. The length of the rectangle is equal to the circumference of the base circle, $2\pi r$. The area for the rectangle can be found by using the formula:

$$A = 2\pi r \times h$$

By adding the two areas for the bases and the area of the rectangle, the surface area of the cylinder can be found, described in units squared.

Finding the Volume and Surface Area of Right Rectangular Prisms, Including Those with Fractional Edge Lengths

Right rectangular prisms are those prisms in which all sides are rectangles and all angles are right, or equal to 90 degrees. The volume for these objects can be found by multiplying the length by the width by the height. The formula is $V = lwh$. For the following prism, the volume formula is:

$$V = 6\frac{1}{2} \times 3 \times 9$$

When dealing with fractional edge lengths, it is helpful to convert the length to an improper fraction. The length 6 ½ cm becomes $\frac{13}{2}$ cm. Then the formula becomes:

$$V = \frac{13}{2} \times 3 \times 9$$

$$\frac{13}{2} \times \frac{3}{1} \times \frac{9}{1} = \frac{351}{2}$$

This value for volume is better understood when turned into a mixed number, which would be 175 ½ cm³.

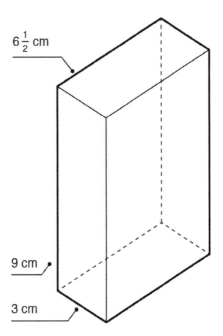

When dimensions for length are given with fractional parts, it can be helpful to turn the mixed number into an improper fraction, then multiply to find the volume, then convert back to a mixed number. When finding surface area, this conversion to improper fractions can also be helpful. The surface area can be found for the same prism above by breaking down the figure into basic shapes. These shapes are rectangles, made up of the two bases, two sides, and the front and back.

The formula for the surface area uses the area for each of these shapes for the terms in the following equation:

$$SA = 6\frac{1}{2} \times 3 + 6\frac{1}{2} \times 3 + 3 \times 9 + 3 \times 9 + 6\frac{1}{2} * 9 + 6\frac{1}{2} \times 9$$

Because there are so many terms in a surface area formula and because this formula contains a fraction, it can be simplified by combining groups that are the same. Each set of numbers is used twice, to represent areas for the opposite sides of the prism. The formula can be simplified to:

$$SA = 2\left(6\frac{1}{2} \times 3\right) + 2(3 \times 9) + 2\left(6\frac{1}{2} \times 9\right)$$

$$2\left(\frac{13}{2} \times 3\right) + 2(27) + 2\left(\frac{13}{2} \times 9\right)$$

$$2\left(\frac{39}{2}\right) + 54 + 2\left(\frac{117}{2}\right)$$

$$39 + 54 + 117 = 210 \text{ cm}^2$$

Understanding Composition of Objects

Composition of objects is the way objects are used in conjunction with each other to form bigger, more complex shapes. For example, a rectangle and a triangle can be used together to form an arrow. Arrows can be found in many everyday scenarios, but are often not seen as the composition of two different shapes. A square is a common shape, but it can also be the composition of shapes. As seen in the second figure, there are many shapes used in the making of the one square. There are five triangles that are three different sizes. There is also one square and one parallelogram used to compose this square. These shapes can be used to compose each more complex shape because they line up, side by side, to fill in the shape with no gaps. This defines composition of shapes where smaller shapes are used to make larger, more complex ones.

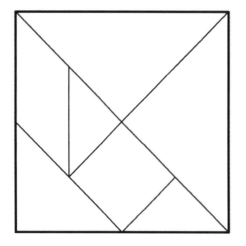
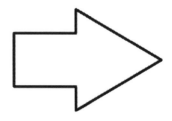

Solving for Missing Values in Triangles, Circles, and Other Figures

Solving for missing values in shapes requires knowledge of the shape and its characteristics. For example, a triangle has three sides and three angles that add up to 180 degrees. If two angle measurements are given, the third can be calculated. For the triangle below, the one given angle has a measure of 55 degrees. The missing angle is *x*. The third angle is labeled with a square, which indicates a measure of 90 degrees. Because all angles must sum to 180 degrees, the following equation can be used to find the missing *x*-value:

$$55 + 90 + x = 180$$

Adding the two given angles and subtracting the total from 180, the missing angle is found to be 35 degrees.

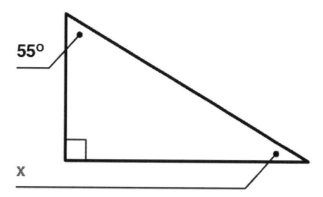

A similar problem can be solved with circles. If the radius is given, but the circumference is unknown, it can be calculated using the formula $C = 2\pi r$. This example can be used in the figure below. The radius can be substituted for r in the formula. Then the circumference can be found as:

$$C = 2\pi \times 8 = 16\pi = 50.24 \ cm$$

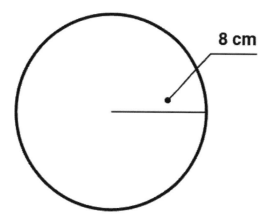

Other figures that may have missing values could be the length of a square, given the area, or the perimeter of a rectangle, given the length and width. All of the missing values can be found by first identifying all the characteristics that are known about the shape, then looking for ways to connect the missing value to the given information.

Within right triangles, trigonometric ratios can be defined for the acute angle within the triangle. Consider the following right triangle. The side across from the right angle is known as the **hypotenuse,** the acute angle being discussed is labeled *θ*, the side across from the acute angle is known as the **opposite** side, and the other side is known as the **adjacent** side.

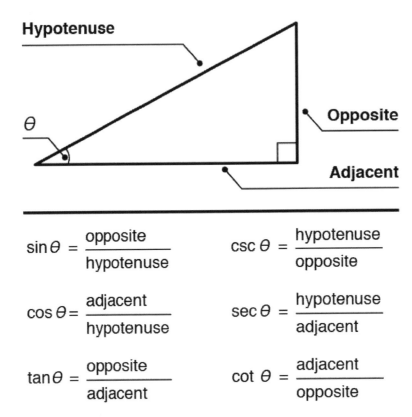

The six trigonometric ratios are shown above as well. "Sin" is short for sine, "cos" is short for cosine, "tan" is short for tangent, "csc" is short for cosecant, "sec" is short for secant, and "cot" is short for cotangent. A mnemonic device exists that is helpful to remember the ratios. SOHCAHTOH stands for Sine = Opposite/Hypotenuse, Cosine = Adjacent/Hypotenuse, and Tangent = Opposite/Adjacent. The other three trigonometric ratios are reciprocals of sine, cosine, and tangent because:

$$\csc \theta = \frac{1}{\sin \theta}, \sec \theta = \frac{1}{\cos \theta}, \text{ and } \cot \theta = \frac{1}{\tan \theta}$$

The **Pythagorean Theorem** is an important relationship between the three sides of a right triangle. It states that the square of hypotenuse is equal to the sum of the squares of the other two sides. When using the Pythagorean Theorem, the hypotenuse is labeled as side c, the opposite is labeled as side *a*, and the adjacent side is side *b*.

The theorem can be seen in the following diagram:

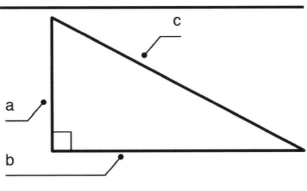

The Pythagorean Theorum

$a^2 + b^2 = c^2$

Both the trigonometric ratios and Pythagorean Theorem can be used in problems that involve finding either a missing side or missing angle of a right triangle. Look to see what sides and angles are given and select the correct relationship that will assist in finding the missing value. These relationships can also be used to solve application problems involving right triangles. Often, it is helpful to draw a figure to represent the problem to see what is missing.

Equations of Conic Sections
The intersection of a plane and a double right circular cone is called a **conic section**. There are four basic types of conic sections, a circle, a parabola, a hyperbola, and an ellipse. The equation of a **circle** is given by:

$$(x - h)^2 + (y - k)^2 = r^2$$

The center of the circle is given by (*h, k*) and the radius of the circle is 5. A parabola that opens up or down has a horizontal axis. The equation of a **parabola with a horizontal axis** is given by:

$$(y - k)^2 = 4p(x - h), \text{ p} \neq 0, \text{ where the vertex is given by } (h, k)$$

A parabola that opens to the left or right has a vertical axis. The equation of the **parabola with a vertical axis** is given by:

$$(x - h)^2 = 4p(y - k), \text{ p} \neq 0, \text{ where the vertex is given by } (h, k)$$

The equation of an **ellipse** with a horizontal major axis is given by:

$$\frac{(x - h)^2}{a^2} + \frac{(y - k)^2}{b^2} = 1$$

with center (*h, k*)

The distance between center and either focus is c with $c^2 = a^2 - b^2$, a > b > 0. The major axis has length $2a$ and the minor axis has length $2b$.

For an ellipse with a vertical major axis, the a and b switch places so the equation is given by:

$$\frac{(x-h)^2}{b^2} + \frac{(y-k)^2}{a^2} = 1$$

a >b > 0, with center (h, k)

The major axis still has length $2a$ and the minor axis still has length $2b$, and the distance between center and either focus is $c^2 = a^2 - b^2$, a > b > 0. A **hyperbola** has an equation similar to the ellipse except that there is a minus in place of the plus sign. A hyperbola with a vertical transverse axis has equation:

$$\frac{(x-h)^2}{a^2} - \frac{(y-k)^2}{b^2} = 1$$

And a hyperbola with a horizontal transverse axis has equation:

$$\frac{(y-k)^2}{a^2} - \frac{(x-h)^2}{b^2} = 1$$

For each of these, the center is given by (h, k) and distance between the vertices $2a$.

Statistics & Probability

Describing Center and Spread of Distributions

One way information can be interpreted from tables, charts, and graphs is through statistics. The three most common calculations for a set of data are the mean, median, and mode. These three are called **measures of central tendency**. Measures of central tendency are helpful in comparing two or more different sets of data.

The **mean** refers to the average and is found by adding up all values and dividing the total by the number of values. In other words, the mean is equal to the sum of all values divided by the number of data entries. For example, if you bowled a total of 532 points in 4 bowling games, your mean score was $\frac{532}{4} = 133$ points per game. A common application of mean useful to students is calculating what he or she needs to receive on a final exam to receive a desired grade in a class.

The **median** is found by lining up values from least to greatest and choosing the middle value. If there's an even number of values, then the mean of the two middle amounts must be calculated to find the median. For example, the median of the set of dollar amounts $5, $6, $9, $12, and $13 is $9. The median of the set of dollar amounts $1, $5, $6, $8, $9, $10 is $7, which is the mean of $6 and $8.

The **mode** is the value that occurs the most. The mode of the data set {1, 3, 1, 5, 5, 8, 10} actually refers to two numbers: 1 and 5. In this case, the data set is bimodal because it has two modes. A data set can have no mode if no amount is repeated. Another useful statistic is range.

The **range** for a set of data refers to the difference between the highest and lowest value.

In some cases, some numbers in a list of data might have weights attached to them. In that case, a **weighted mean** can be calculated. A common application of a weighted mean is GPA. In a semester, each class is assigned a number of credit hours, its weight, and at the end of the semester each student receives a grade. To compute GPA, an A is a 4, a B is a 3, a C is a 2, a D is a 1, and an F is a 0. Consider a student that takes a 4-hour English class, a 3-hour math class, and a 4-hour history class and receives all B's. The weighted mean, GPA, is found by multiplying each grade times its weight, number of credit hours, and dividing by the total number of credit hours.

Therefore, the student's GPA is:

$$\frac{3 \times 4 + 3 \times 3 + 3 \times 4}{11} = \frac{33}{1} = 3.0.$$

The following bar chart shows how many students attend a cycle class on each day of the week. To find the mean attendance for the week, each day's attendance can be added together, $10 + 7 + 6 + 9 + 8 + 14 + 4 = 58$, and the total divided by the number of days, $58 \div 7 = 8.3$. The mean attendance for the week was 8.3 people. The median attendance can be found by putting the attendance numbers in order from least to greatest: 4, 6, 7, 8, 9, 10, 14, and choosing the middle number: 8 people. The mode for attendance is none for this set of data because no numbers repeat. The range is 10, which is found by finding the difference between the lowest number, 4, and the highest number, 14.

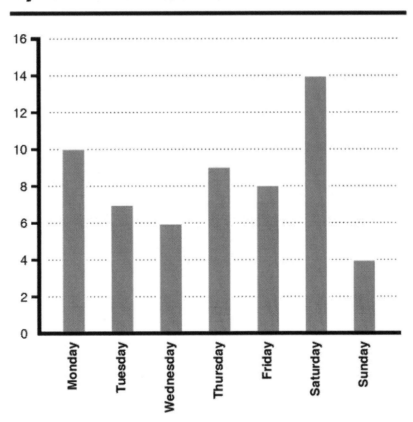

Cycle class attendance

A **histogram** is a bar graph used to group data into "bins" that cover a range on the horizontal, or x-axis. Histograms consist of rectangles whose height is equal to the frequency of a specific category. The horizontal axis represents the specific categories. Because they cover a range of data, these bins have no gaps between bars, unlike the bar graph above. In a histogram showing the heights of adult golden retrievers, the bottom axis would be groups of heights, and the y-axis would be the number of dogs in each range. Evaluating this histogram would show the height of most golden retrievers as falling within a certain range. It also provides information to find the average height and range for how tall golden retrievers may grow.

The following is a histogram that represents exam grades in a given class. The horizontal axis represents ranges of the number of points scored, and the vertical axis represents the number of students. For example, approximately 33 students scored in the 60 to 70 range.

Results of the exam

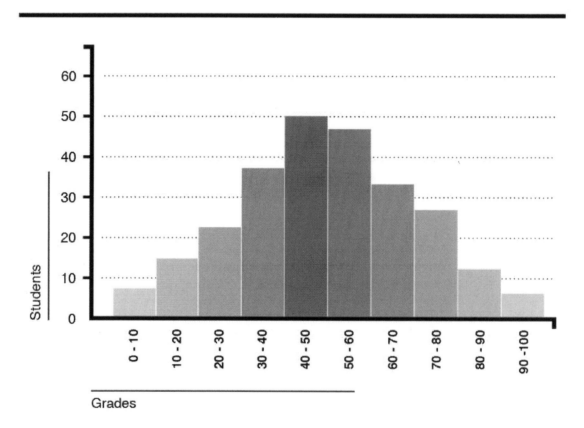

Certain measures of central tendency can be easily visualized with a histogram. If the points scored were shown with individual rectangles, the tallest rectangle would represent the mode. A bimodal set of data would have two peaks of equal height. Histograms can be classified as having data **skewed to the left, skewed to the right,** or **normally distributed**, which is also known as **bell-shaped**. These three classifications can be seen in the following chart:

Measures of central tendency images

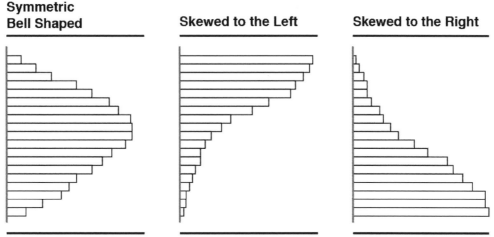

When the data is normal, the mean, median, and mode are all very close. They all represent the most typical value in the data set. The mean is typically used as the best measure of central tendency in this case because it does include all data points. However, if the data is skewed, the mean becomes less meaningful. The median is the best measure of central tendency because it is not affected by any outliers, unlike the mean. When the data is skewed, the mean is dragged in the direction of the skew. Therefore, if the data is not normal, it is best to use the median as the measure of central tendency.

The measures of central tendency and the range may also be found by evaluating information on a line graph.

The line graph shows the daily high and low temperatures:

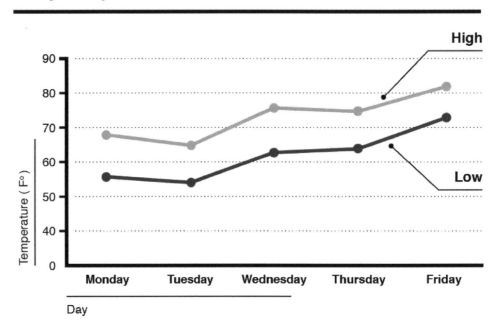

Daily Temperatures

The average high temperature can be found by gathering data from each day on the triangle line. The days' highs are 82, 78, 75, 65, and 70. The average is found by adding them together to get 370, then dividing by 5 (because there are 5 temperatures). The average high for the five days is 74. If 74 degrees is found on the graph, then it falls in the middle of the values on the triangle line. The average low temperature can be found in the same way.

Given a set of data, the **correlation coefficient**, r, measures the association between all the data points. If two values are correlated, there is an association between them. However, correlation does not necessarily mean causation, or that that one value causes the other. There is a common mistake made that assumes correlation implies causation. Average daily temperature and number of sunbathers are both correlated and have causation. If the temperature increases, that change in weather causes more people to want to catch some rays. However, wearing plus-size clothing and having heart disease are two variables that are correlated but do not have causation. The larger someone is, the more likely he or she is to have heart disease. However, being overweight does not cause someone to have the disease.

The value of the correlation coefficient is between −1 and 1, where −1 represents a perfect negative linear relationship, 0 represents no relationship between the two data sets, and 1 represents a perfect positive linear relationship. A negative linear relationship means that as x-values increase, y-values decrease. A positive linear relationship means that as x values increase, y values increase.

The formula for computing the correlation coefficient is:

$$r = \frac{n \sum xy - (\sum x)(\sum y)}{\sqrt{n(\sum x^2) - (\sum x)^2}\sqrt{n(\sum y^2) - (y)^2}}$$

where n is the number of data points. The closer r is to 1 or -1, the stronger the correlation. A correlation can be seen when plotting data. If the graph resembles a straight line, there is a correlation.

Applying and Analyzing Data Collection Methods

Data collection can be done through surveys, experiments, observations, and interviews. A **census** is a type of survey that is done with a whole population. Because it can be difficult to collect data for an entire population, sometimes, a sample survey is used. In this case, one would survey only a fraction of the population and make inferences about the data and generalizations about the larger population from which the sample was drawn. Sample surveys are not as accurate as a census, but this is an easier and less expensive method of collecting data. An **experiment** is used when a researcher wants to explain how one variable causes changes in another variable. For example, if a researcher wanted to know if a particular drug affects weight loss, he or she would choose a treatment group that would take the drug, and another group, the control group, that would not take the drug.

Special care must be taken when choosing these groups to ensure that bias is not a factor. **Bias** occurs when an outside factor influences the outcome of the research. In observational studies, the researcher does not try to influence either variable, but simply observes the behavior of the subjects. Interviews are sometimes used to collect data as well. The researcher will ask questions that focus on her area of interest in order to gain insight from the participants. When gathering data through observation or interviews, it is important that the researcher be well trained so that he or she does not influence the results and so that the study is reliable. A study is reliable if it can be repeated under the same conditions and the same results are received each time.

Understanding and Modeling Relationships in Bivariate Data

Independent and dependent are two types of variables that describe how they relate to each other. The **independent variable** is the variable controlled by the experimenter. It stands alone and isn't changed by other parts of the experiment. This variable is normally represented by x and is found on the horizontal, or x-axis, of a graph. The **dependent variable** changes in response to the independent variable. It reacts to, or depends on, the independent variable. This variable is normally represented by y and is found on the vertical, or y-axis of the graph.

The relationship between two variables, x and y, can be seen on a scatterplot.

The following scatterplot shows the relationship between weight and height. The graph shows the weight as *x* and the height as *y*. The first dot on the left represents a person who is 45 kg and approximately 150 cm tall. The other dots correspond in the same way. As the dots move to the right and weight increases, height also increases. A line could be drawn through the middle of the dots to move from bottom left to top right. This line would indicate a **positive correlation** between the variables. If the variables had a **negative correlation**, then the dots would move from the top left to the bottom right.

Height and Weight

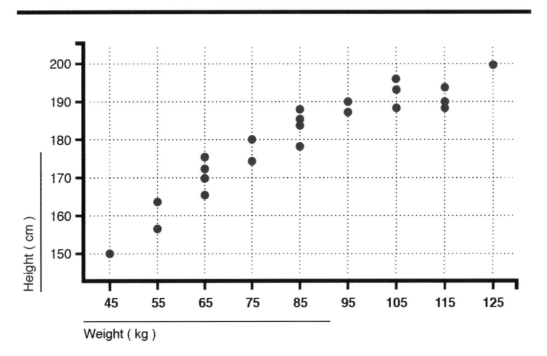

A **scatterplot** is useful in determining the relationship between two variables, but it's not required. Consider an example where a student scores a different grade on his math test for each week of the month. The independent variable would be the weeks of the month. The dependent variable would be the grades, because they change depending on the week. If the grades trended up as the weeks passed, then the relationship between grades and time would be positive. If the grades decreased as the time passed, then the relationship would be negative. (As the number of weeks went up, the grades went down.)

The relationship between two variables can further be described as strong or weak. The relationship between age and height shows a strong positive correlation because children grow taller as they get older. In adulthood, the relationship between age and height becomes weak, and the dots will spread out. People stop growing in adulthood, and their final heights vary depending on factors like genetics and health. The closer the dots on the graph, the stronger the relationship. As they spread apart, the relationship becomes weaker. If they are too spread out to determine a correlation up or down, then the variables are said to have no correlation.

Variables are values that change, so determining the relationship between them requires an evaluation of who or what changes them. If the variable changes because of a result in the experiment, then it's dependent. If the variable changes before the experiment, or is changed by the person controlling the experiment, then it's the independent variable. As they interact, one is manipulated by the other. The manipulator is the independent, and the manipulated is the dependent. Once the independent and dependent variable are determined, they can be evaluated to have a positive, negative, or no correlation.

Calculating Probabilities

Probability describes how likely it is that an event will occur. Probabilities are always a number from 0 to 1. If an event has a high likelihood of occurrence, it will have a probability close to 1. If there is only a small chance that an event will occur, the likelihood is close to 0. A fair six-sided die has one of the numbers 1, 2, 3, 4, 5, and 6 on each side. When this die is rolled there is a one in six chance that it will land on 2. This is because there are six possibilities and only one side has a 2 on it. The probability then is $\frac{1}{6}$ or .167. The probability of rolling an even number from this die is three in six, or ½ or .5. This is because there are three sides on the die with even numbers (2, 4, 6), and there are six possible sides. The probability of rolling a number less than 10 is one because every side of the die has a number less than 6, so this is certain to occur. On the other hand, the probability of rolling a number larger than 20 is zero. There are no numbers greater than 20 on the die, so it is certain that this will not occur; thus, the probability is zero.

If a teacher says that the probability of anyone passing her final exam is .2, is it highly likely that anyone will pass? No, the probability of anyone passing her exam is low because .2 is closer to 0 than to 1. If another teacher is proud that the probability of students passing his class is .95, how likely is it that a student will pass? It is highly likely that a student will pass because the probability, .95, is very close to 1.

Using Two-Way Tables to Summarize Categorical Data and Relative Frequencies, and to Calculate Conditional Probability

A **two-way frequency table** displays categorical data with two variables, and it highlights relationships that exist between those two variables. Such tables are used frequently to summarize survey results, and are also known as **contingency tables**. Each cell shows a count pertaining to that individual variable paring, known as a **joint frequency**, and the totals of each row and column also are in the table. Consider the following two-way frequency table:

Distribution of the Residents of a Particular Village

	70 or older	69 or younger	Totals
Women	20	40	60
Men	5	35	40
Total	25	75	100

The table above shows the breakdown of ages and sexes of 100 people in a particular village. The total number of people in the data is shown in the bottom right corner. Each total is shown at the end of each row or column, as well. For instance, there were 25 people age 70 or older and 60 women in the data. The 20 in the first cell shows that out of 100 total villagers, 20 were women aged 70 or older. The 5 in the cell below shows that out of 100 total villagers, 5 were men aged 70 or older.

A two-way table can also show relative frequencies. If instead of the count, the percentage of people in each category was placed into the cells, the two-way table would show relative frequencies. If each frequency is calculated over the entire total of 100, the first cell would be 20% or 0.2. However, the relative frequencies can also be calculated over row or column totals. If row totals were used, the first cell would be:

$$\frac{20}{60} = 0.333 = 33.3\%$$

If column totals were used, the first cell would be:

$$\frac{20}{25} = 0.8 = 80\%$$

Such tables can be used to calculate **conditional probabilities**, which are probabilities that an event occurs, given another event. Consider a randomly selected villager. The probability of selecting a male 70 years old or older is $\frac{5}{100} = 0.05$ because there are 5 males over the age of 70 and 100 total villagers.

Integrating Essential Skills

Rates and Percentages

Ratios and Rates of Change

Recall that a **ratio** is the comparison of two different quantities. Comparing 2 apples to 3 oranges results in the ratio 2:3, which can be expressed as the fraction $\frac{2}{3}$. Note that order is important when discussing ratios. The number mentioned first is the numerator, and the number mentioned second is the denominator. The ratio 2:3 does not mean the same quantitative relationship as the ratio 3:2. Also, it is important to make sure than when discussing ratios that have units attached to them, the two quantities use the same units. For example, to think of 8 feet to 4 yards, it would make sense to convert 4 yards to feet by multiplying by 3. Therefore, the ratio would be 8 feet to 12 feet, which can be expressed as the fraction $\frac{8}{12}$. Also, note that it is proper to refer to ratios in lowest terms.

Therefore, the ratio of 8 feet to 4 yards is equivalent to the fraction $\frac{2}{3}$. Many real-world problems involve ratios. Often, problems with ratios involve proportions, as when two ratios are set equal to find the missing amount. However, some problems involve deciphering single ratios. For example, consider an amusement park that sold 345 tickets last Saturday. If 145 tickets were sold to adults and the rest of the tickets were sold to children, what would the ratio of the number of adult tickets to children's tickets be? A common mistake would be to say the ratio is 145:345. However, 345 is the total number of tickets sold. There were 345 − 145 = 200 tickets sold to children. Thus, the correct ratio of adult to children's tickets is 145:200. As a fraction, this expression is written as $\frac{145}{200}$, which can be reduced to $\frac{29}{40}$.

While a ratio compares two measurements using the same units, **rates** compare two measurements with different units. Examples of rates would be $200 of work for 8 hours, or 500 miles per 20 gallons. Because the units are different, it is important to always include the units when discussing rates. Rates can be easily seen because if they are expressed in words, the two quantities are usually split up using one of the following words: *for, per, on, from, in*. Just as with ratios, it is important to write rates in lowest terms. A common rate that can be found in many real-life situations is cost per unit. This quantity describes how much one item or one unit costs. This rate allows the best buy to be determined, given a couple of different sizes of an item with different costs. For example, if 2 quarts of soup was sold for $3.50 and 3 quarts was sold for $4.60, to determine the best buy, the cost per quart should be found. $\frac{\$3.50}{2} = \$1.75$ per quart, and $\frac{\$4.60}{3} = \$1.53$ per quart. Therefore, the better deal would be the 3-quart option.

Rate of change problems involve calculating a quantity per some unit of measurement. Usually the unit of measurement is time. For example, meters per second is a common rate of change. To calculate this measurement, find the distance traveled in meters and divide by total time traveled. The calculation is an average of the speed over the entire time interval. Another common rate of change used in the real world is miles per hour. Consider the following problem that involves calculating an average rate of change in temperature. Last Saturday, the temperature at 1:00 a.m. was 34 degrees Fahrenheit, and at noon, the temperature had increased to 75 degrees Fahrenheit. What was the average rate of change over that time interval? The average rate of change is calculated by finding the total change in temperature and dividing by the total hours elapsed. Therefore, the rate of change was equal to:

$$\frac{75-34}{12-1} = \frac{41}{11} \text{ degrees per hour}$$

This quantity, rounded to two decimal places, is equal to 3.72 degrees per hour.

A common rate of change that appears in algebra is the slope calculation. Given a linear equation in one variable, $y = mx + b$, the slope, m, is equal to $\frac{rise}{run}$ or $\frac{change\ in\ y}{change\ in\ x}$. In other words, slope is equivalent to the ratio of the vertical and horizontal changes between any two points on a line. The vertical change is known as the **rise**, and the horizontal change is known as the **run**. Given any two points on a line (x_1, y_1) and (x_2, y_2), slope can be calculated with the formula:

$$m = \frac{y_2 - y_1}{x_2 - x_1} = \frac{\Delta y}{\Delta x}$$

Common real-world applications of slope include determining how steep a staircase should be, calculating how steep a road is, and determining how to build a wheelchair ramp.

Many times, problems involving rates and ratios involve proportions. A **proportion** states that two ratios (or rates) are equal. The property of cross products can be used to determine if a proportion is true, meaning both ratios are equivalent. If $\frac{a}{b} = \frac{c}{d}$, then to clear the fractions, multiply both sides by the least common denominator, bd. This results in $ac = cd$, which is equal to the result of multiplying along both diagonals. For example, $\frac{4}{40} = \frac{1}{10}$ grants the cross product $4 \times 10 = 40 \times 1$. $40 = 40$ shows that this proportion is true. Cross products are used when proportions are involved in real-world problems. Consider the following: If 3 pounds of fertilizer will cover 75 square feet of grass, how many pounds are needed for 375 square feet? To solve this problem, a proportion can be set up using two ratios. Let x equal the unknown quantity, pounds needed for 375 feet. Then, the equation found by setting the two given ratios equal to one another is:

$$\frac{3}{75} = \frac{x}{375}$$

Cross multiplication gives:

$$3 \times 375 = 75x$$

Therefore, $1{,}125 = 75x$. Divide both sides by 75 to get $x = 15$. Therefore, 15 gallons of fertilizer is needed to cover 75 square feet of grass.

Another application of proportions involves similar triangles. If two triangles have the same measurement as two triangles in another triangle, the triangles are said to be **similar.** If two are the same, the third pair of angles are equal as well because the sum of all angles in a triangle is equal to 180 degrees. Each pair of equivalent angles are known as **corresponding angles. Corresponding sides** face the corresponding angles, and it is true that corresponding sides are in proportion.

For example, consider the following set of similar triangles:

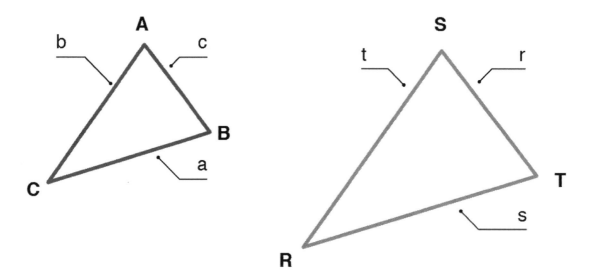

Angles A and R have the same measurement, angles C and T have the same measurement, and angles B and S have the same measurement. Therefore, the following proportion can be set up from the sides:

$$\frac{c}{t} = \frac{a}{r} = \frac{b}{s}$$

This proportion can be helpful in finding missing lengths in pairs of similar triangles. For example, if the following triangles are similar, a proportion can be used to find the missing side lengths, *a* and *b*.

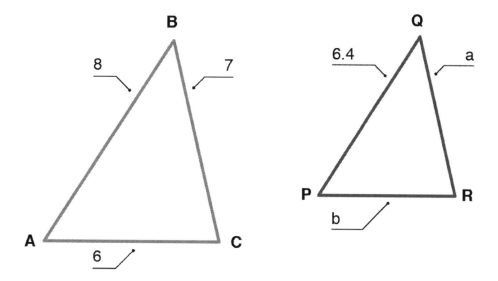

The proportions $\frac{8}{6.4} = \frac{6}{b}$ and $\frac{8}{6.4} = \frac{7}{a}$ can both be cross multiplied and solved to obtain *a* = 5.6 and *b* = 4.8.

A real-life situation that uses similar triangles involves measuring shadows to find heights of unknown objects. Consider the following problem: A building casts a shadow that is 120 feet long, and at the same time, another building that is 80 feet high casts a shadow that is 60 feet long. How tall is the first building? Each building, together with the sun rays and shadows casted on the ground, forms a triangle. They are similar because each building forms a right angle with the ground, and the sun rays form equivalent angles. Therefore, these two pairs of angles are both equal. Because all angles in a triangle add up to 180 degrees, the third angles are equal as well. Both shadows form corresponding sides of the triangle, the buildings form corresponding sides, and the sun rays form corresponding sides. Therefore, the triangles are similar, and the following proportion can be used to find the missing building length:

$$\frac{120}{x} = \frac{60}{80}$$

Cross-multiply to obtain the cross products, $9600 = 60x$. Then, divide both sides by 60 to obtain $x = 160$. This solution means that the other building is 160 feet high.

Percentages
Percentages are defined to be parts per one hundred. To convert a decimal to a percentage, move the decimal point two units to the right and place the percent sign after the number. Percentages appear in many scenarios in the real world. It is important to make sure the statement containing the percentage is translated to a correct mathematical expression. Be aware that it is extremely common to make a mistake when working with percentages within word problems.

An example of a word problem containing a percentage is the following: 35% of people speed when driving to work. In a group of 5,600 commuters, how many would be expected to speed on the way to their place of employment? The answer to this problem is found by finding 35% of 5,600. First, change the percentage to the decimal 0.35. Then compute the product: $0.35 \times 5,600 = 1,960$. Therefore, it would be expected that 1,960 of those commuters would speed on their way to work based on the data given. In this situation, the word "of" signals to use multiplication to find the answer. Another way percentages are used is in the following problem: Teachers work 8 months out of the year. What percent of the year do they work? To answer this problem, find what percent of 12 the number 8 is, because there are 12 months in a year. Therefore, divide 8 by 12, and convert that number to a percentage:

$$\frac{8}{12} = \frac{2}{3} = 0.66\overline{6}$$

The percentage rounded to the nearest tenth place tells us that teachers work 66.7% of the year. Percentages also appear in real-world application problems involving finding missing quantities like in the following question: 60% of what number is 75? To find the missing quantity, an equation can be used. Let x be equal to the missing quantity. Therefore, $0.60x = 75$. Divide each side by 0.60 to obtain 125. Therefore, 60% of 125 is equal to 75.

Sales tax is an important application relating to percentages because tax rates are usually given as percentages. For example, a city might have an 8% sales tax rate. Therefore, when an item is purchased with that tax rate, the real cost to the customer is 1.08 times the price in the store. For example, a $25 pair of jeans costs the customer:

$$\$25 \times 1.08 = \$27$$

Sales tax rates can also be determined if they are unknown when an item is purchased. If a customer visits a store and purchases an item for $21.44, but the price in the store was $19, they can find the tax rate by first subtracting $21.44 − $19 to obtain $2.44, the sales tax amount. The sales tax is a percentage of the in-store price. Therefore, the tax rate is $\frac{2.44}{19} = 0.128$, which has been rounded to the nearest thousandths place. In this scenario, the actual sales tax rate given as a percentage is 12.8%.

Solving Unit Rate Problems
A **unit rate** is a rate with a denominator of one. It is a comparison of two values with different units where one value is equal to one. Examples of unit rates include 60 miles per hour and 200 words per minute. Problems involving unit rates may require some work to find the unit rate. For example, if Mary travels 360 miles in 5 hours, what is her speed, expressed as a unit rate? The rate can be expressed as the following fraction:

$$\frac{360 \; miles}{5 \; hours}$$

The denominator can be changed to one by dividing by five. The numerator will also need to be divided by five to follow the rules of equality. This division turns the fraction into $\frac{72 \; miles}{1 \; hour}$, which can now be labeled as a unit rate because one unit has a value of one. Another type question involves the use of unit rates to solve problems. For example, if Trey needs to read 300 pages and his average speed is 75 pages per hour, will he be able to finish the reading in 5 hours? The unit rate is 75 pages per hour, so the total of 300 pages can be divided by 75 to find the time. After the division, the time it takes to read is four hours. The answer to the question is yes, Trey will finish the reading within 5 hours.

Proportional Relationships

Fractions appear in everyday situations, and in many scenarios, they appear in the real-world as ratios and in proportions. A **ratio** is formed when two different quantities are compared. For example, in a group of 50 people, if there are 33 females and 17 males, the ratio of females to males is 33 to 17. This expression can be written in the fraction form, $\frac{33}{17}$, or by using the ratio symbol, 33:17. The order of the number matters when forming ratios. In the same setting, the ratio of males to females is 17 to 33, which is equivalent to $\frac{17}{33}$ or 17:33. A **proportion** is an equation involving two ratios. The equation $\frac{a}{b} = \frac{c}{d}$, or $a:b = c:d$ is a proportion, for real numbers a, b, c, and d. Usually, in one ratio, one of the quantities is unknown, and cross-multiplication is used to solve for the unknown. Consider $\frac{1}{4} = \frac{x}{5}$. To solve for x, cross-multiply to obtain $5 = 4x$. Divide each side by 4 to obtain the solution $x = \frac{5}{4}$. It is also true that percentages are ratios in which the second term is 100. For example, 65% is 65:100 or $\frac{65}{100}$. Therefore, when working with percentages, one is also working with ratios.

Real-world problems frequently involve proportions. For example, consider the following problem: If 2 out of 50 pizzas are usually delivered late from a local Italian restaurant, how many would be late out of 235 orders? The following proportion would be solved with x as the unknown quantity of late pizzas:

$$\frac{2}{50} = \frac{x}{235}$$

Cross multiplying results in $470 = 50x$. Divide both sides by 50 to obtain $x = \frac{470}{50}$, which in lowest terms is equal to $\frac{47}{5}$. In decimal form, this improper fraction is equal to 9.4. Because it does not make sense to answer this question with decimals (portions of pizzas do not get delivered) the answer must be rounded. Traditional rounding rules would say that 9 pizzas would be expected to be delivered late. However, to be safe, rounding up to 10 pizzas out of 235 would probably make more sense.

Area, Surface Area, and Volume

Perimeter and area are two commonly used geometric quantities that describe objects. **Perimeter** is the distance around an object. The perimeter of an object can be found by adding the lengths of all sides. Perimeter may be used in problems dealing with lengths around objects such as fences or borders. It may also be used in finding missing lengths, or working backwards. If the perimeter is given, but a length is missing, use subtraction to find the missing length.

Given a square with side length s, the formula for perimeter is $P = 4s$. Given a rectangle with length l and width w, the formula for perimeter is $P = 2l + 2w$. The perimeter of a triangle is found by adding the three side lengths, and the perimeter of a trapezoid is found by adding the four side lengths. The units for perimeter are always the original units of length, such as meters, inches, miles, etc. When discussing a circle, the distance around the object is referred to as its **circumference,** not perimeter. The formula for circumference of a circle is $C = 2\pi r$, where r represents the radius of the circle. This formula can also be written as $C = d\mu$, where d represents the diameter of the circle.

Area is the two-dimensional space covered by an object. These problems may include the area of a rectangle, a yard, or a wall to be painted. Finding the area may be a simple formula, or it may require multiple formulas to be used together. The units for area are square units, such as square meters, square inches, and square miles. Given a square with side length s, the formula for its area is $A = s^2$.

Some other common shapes are shown below:

Shape	Formula	Graphic
Rectangle	$Area = length \times width$	
Triangle	$Area = \dfrac{1}{2} \times base \times height$	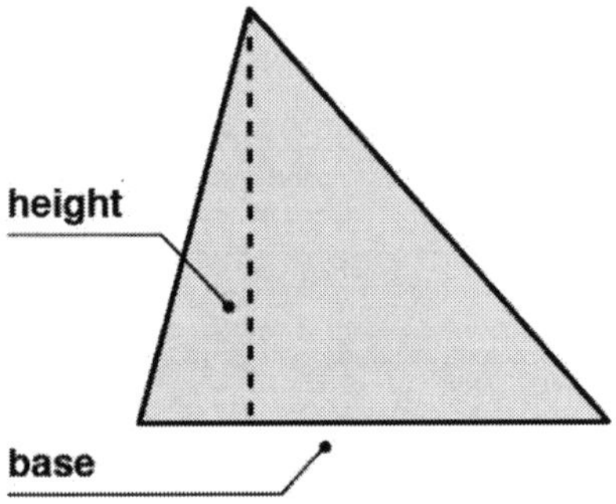 **height** **base**
Circle	$Area = \pi \times radius^2$	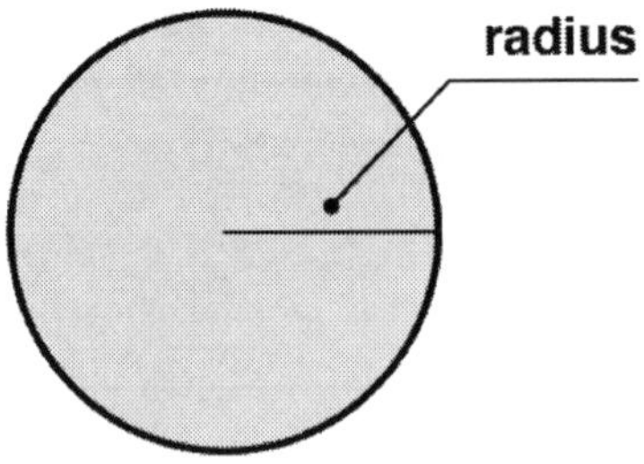 **radius**

The following formula, not as widely used as those shown above, but very important, is the area of a trapezoid:

Area of a Trapezoid

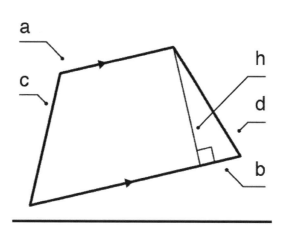

$$A = \frac{1}{2}(a+b)h$$

To find the area of the shapes above, use the given dimensions of the shape in the formula. Complex shapes might require more than one formula. To find the area of the figure below, break the figure into two shapes. The rectangle has dimensions 6 cm by 7 cm. The triangle has dimensions 6 cm by 6 cm. Plug the dimensions into the rectangle formula:

$$A = 6 \times 7$$

Multiplication yields an area of 42 cm². The triangle area can be found using the formula:

$$A = \frac{1}{2} \times 4 \times 6$$

Multiplication yields an area of 12 cm².

Add the areas of the two shapes to find the total area of the figure, which is 54 cm².

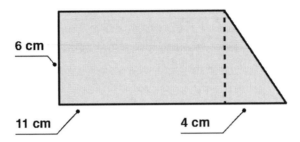

Instead of combining areas, some problems may require subtracting them, or finding the difference.

To find the area of the shaded region in the figure below, determine the area of the whole figure. Then the area of the circle can be subtracted from the whole.

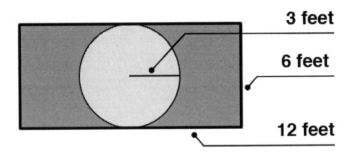

The following formula shows the area of the outside rectangle:

$$A = 12 \times 6 = 72 \ ft^2$$

The area of the inside circle can be found by the following formula:

$$A = \pi(3)^2 = 9\pi = 28.3 \ ft^2$$

As the shaded area is outside the circle, the area for the circle can be subtracted from the area of the rectangle to yield an area of $43.7 \ ft^2$.

While some geometric figures may be given as pictures, others may be described in words. If a rectangular playing field with dimensions 95 meters long by 50 meters wide is measured for perimeter, the distance around the field must be found. The perimeter includes two lengths and two widths to measure the entire outside of the field. This quantity can be calculated using the following equation:

$$P = 2(95) + 2(50) = 290 \ m$$

The distance around the field is 290 meters.

Perimeter and area are two-dimensional descriptions; volume is three-dimensional. **Volume** describes the amount of space that an object occupies, but it's different from area because it has three dimensions instead of two. The units for volume are cubic units, such as cubic meters, cubic inches, and cubic miles. Volume can be found by using formulas for common objects such as cylinders and boxes.

The following chart shows a diagram and formula for the volume of two objects.

Shape	Formula	Diagram
Rectangular Prism (box)	$V = length \times width \times height$	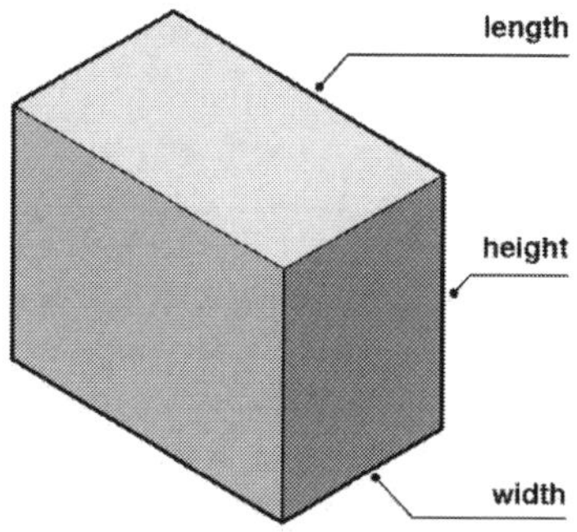
Cylinder	$V = \pi \times radius^2 \times height$	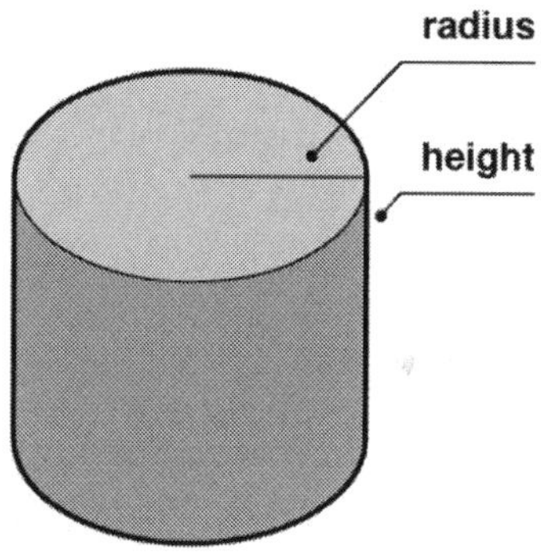

Volume formulas of these two objects are derived by finding the area of the bottom two-dimensional shape, such as the circle or rectangle, and then multiplying times the height of the three-dimensional shape. Other volume formulas include the volume of a cube with side length s: $V = s^3$; the volume of a sphere with radius r: $V = \frac{4}{3}\pi r^3$; and the volume of a cone with radius r and height h:

$$V = \frac{1}{3}\pi r^2 h$$

If a soda can has a height of 5 inches and a radius on the top of 1.5 inches, the volume can be found using one of the given formulas. A soda can is a cylinder. Knowing the given dimensions, the formula can be completed as follows:

$$V = \pi \, (radius)^2 \times height$$

$$\pi \, (1.5)^2 \times 5 = 35.325 \; inches^3$$

Notice that the units for volume are inches cubed because it refers to the number of cubic inches required to fill the can.

With any geometric calculations, it's important to determine what dimensions are given and what quantities the problem is asking for. If a connection can be made between them, the answer can be found.

Other geometric quantities can include angles inside a triangle. The sum of the measures of three angles in any triangle is 180 degrees. Therefore, if only two angles are known inside a triangle, the third can be found by subtracting the sum of the two known quantities from 180. Two angles whose sum is equal to 90 degrees are known as **complementary angles.** For example, angles measuring 72 and 18 degrees are complementary, and each angle is a complement of the other. Finally, two angles whose sum is equal to 180 degrees are known as **supplementary angles.** To find the supplement of an angle, subtract the given angle from 180 degrees. For example, the supplement of an angle that is 50 degrees is

$$180 - 50 = 130 \text{ degrees}$$

These terms involving angles can be seen in many types of word problems. For example, consider the following problem: The measure of an angle is 60 degrees less than two times the measure of its complement. What is the angle's measure? To solve this, let x be the unknown angle. Therefore, its complement is $90 - x$. The problem gives that:

$$x = 2(90 - x) - 60$$

To solve for x, distribute the 2, and collect like terms. This process results in:

$$x = 120 - 2x$$

Then, use the addition property to add $2x$ to both sides to obtain:

$$3x = 120$$

Finally, use the multiplication properties of equality to divide both sides by 3 to get:

$$x = 40$$

Therefore, the angle measures 40 degrees. Also, its complement measures 50 degrees.

Determining Surface Area Measurements
As mentioned previously, **surface area** is defined as the area of the surface of a figure. A pyramid has a surface made up of four triangles and one square. To calculate the surface area of a pyramid, the areas of each individual shape are calculated. Then the areas are added together. This method of decomposing the shape into two-dimensional figures to find area, then adding the areas, can be used to find surface area for any figure. Once these measurements are found, the area is described with square units. The surface area of a cube is found by multiplying the area of one side by six since there are six equivalent sides. For example, if the side length is 4 inches, the surface area is found using $SA = 6 \times (4 \times 4) = 96 \; square \; inches.$

Average and Median

A data set can be described by calculating the mean, median, and mode. These values, called **measures of center,** allow the data to be described with a single value that is representative of the data set.

Again, the most common measure of center is the **mean,** also referred to as the **average.**

To calculate the mean,

- Add all data values together

- Divide by the sample size (the number of data points in the set)

The **median** is middle data value, so that half of the data lies below this value and half lies below the data value.

To calculate the median,

- Order the data from least to greatest

- The point in the middle of the set is the median

  - In the event that there is an even number of data points, add the two middle points and divide by 2

The **mode** is the data value that occurs most often.

To calculate the mode,

- Order the data from least to greatest

- Find the value that occurs most often

Example: Amelia is a leading scorer on the school's basketball team. The following data set represents the number of points that Amelia has scored in each game this season. Use the mean, median, and mode to describe the data.

16, 12, 26, 14, 13, 28, 14, 12, 15, 25

Solution:

Mean:

$$16 + 12 + 26 + 14 + 28 + 14 + 12 + 15 + 25 = 162$$

$$162 \div 9 = 18$$

Amelia averages 18 points per game.

Median:

12, 12, 14, 14, **15**, 16, 25, 26, 28

Amelia's median score is 15.

Mode:

12, 12, 14, 14, 15, 16, 25, 26, 28

The numbers 12 and 14 each occur twice in the data set, so this set has 2 modes: 12 and 14.

The **range** is the difference between the largest and smallest values in the set. In the example above, the range is:

$$28 - 12 = 16$$

Expressing Numbers in Different Ways

Converting Non-Negative Fractions, Decimals, and Percentages

Within the number system, different forms of numbers can be used. It is important to be able to recognize each type, as well as work with, and convert between, the given forms. The **real number system** comprises natural numbers, whole numbers, integers, rational numbers, and irrational numbers. Natural numbers, whole numbers, integers, and irrational numbers typically are not represented as fractions, decimals, or percentages. Rational numbers, however, can be represented as any of these three forms. A **rational number** is a number that can be written in the form $\frac{a}{b}$, where a and b are integers, and b is not equal to zero. In other words, rational numbers can be written in a fraction form. The value a is the **numerator**, and b is the **denominator.**

If the numerator is equal to zero, the entire fraction is equal to zero. Non-negative fractions can be less than 1, equal to 1, or greater than 1. Fractions are less than 1 if the numerator is smaller (less than) than the denominator. For example, $\frac{3}{4}$ is less than 1. A fraction is equal to 1 if the numerator is equal to the denominator. For instance, $\frac{4}{4}$ is equal to 1. Finally, a fraction is greater than 1 if the numerator is greater than the denominator: the fraction $\frac{11}{4}$ is greater than 1.

When the numerator is greater than the denominator, the fraction is called an **improper fraction**. An improper fraction can be converted to a **mixed number,** a combination of both a whole number and a fraction. To convert an improper fraction to a mixed number, divide the numerator by the denominator. Write down the whole number portion, and then write any remainder over the original denominator. For example, $\frac{11}{4}$ is equivalent to $2\frac{3}{4}$. Conversely, a mixed number can be converted to an improper fraction by multiplying the denominator by the whole number and adding that result to the numerator.

Fractions can be converted to decimals. With a calculator, a fraction is converted to a decimal by dividing the numerator by the denominator. For example:

$$\frac{2}{5} = 2 \div 5 = 0.4$$

Sometimes, rounding might be necessary. Consider:

$$\frac{2}{7} = 2 \div 7 = 0.28571429$$

This decimal could be rounded for ease of use, and if it needed to be rounded to the nearest thousandth, the result would be 0.286. If a calculator is not available, a fraction can be converted to a decimal manually. First, find a number that, when multiplied by the denominator, has a value equal to 10, 100, 1,000, etc. Then, multiply both the numerator and denominator times that number. The decimal form of the fraction is equal to the new numerator with a decimal point placed as many place values to the left as there are zeros in the denominator. For example, to convert $\frac{3}{5}$ to a decimal, multiply both the numerator and denominator times 2, which results in $\frac{6}{10}$. The decimal is equal to 0.6 because there is one zero in the denominator, and so the decimal place in the numerator is moved one unit to the left. In the case where rounding would be necessary while working without a calculator, an approximation must be found. A number close to 10, 100, 1,000, etc. can be used. For example, to convert $\frac{1}{3}$ to a decimal, the numerator and denominator can be multiplied by 33 to turn the denominator into approximately 100, which makes for an easier conversion to the equivalent decimal. This process results in $\frac{33}{99}$ and an approximate decimal of 0.33. Once in decimal form, the number can be converted to a percentage. Multiply the decimal by 100 and then place a percent sign after the number. For example, .614 is equal to 61.4%. In other words, move the decimal place two units to the right and add the percentage symbol.

Composing and Decomposing Multidigit Numbers

Composing and decomposing numbers reveals the place value held by each number 0 through 9 in each position. For example, the number 17 is read as "seventeen." It can be decomposed into the numbers 10 and 7. It can be described as 1 group of ten and 7 ones. The one in the tens place represents one set of ten. The seven in the ones place represents seven sets of one. Added together, they make a total of seventeen. The number 48 can be written in words as "forty-eight." It can be decomposed into the numbers 40 and 8, where there are 4 groups of ten and 8 groups of one. The number 296 can be decomposed into 2 groups of one hundred, 9 groups of ten, and 6 groups of one. There are two hundreds, nine tens, and six ones. Decomposing and composing numbers lays the foundation for visually picturing the number and its place value, and adding and subtracting multiple numbers with ease.

Modeling

Producing, Interpreting, Understanding, Evaluating, and Improving Models

Interpreting Relevant Information from Tables, Charts, and Graphs

Tables, charts, and graphs can be used to convey information about different variables. They are all used to organize, categorize, and compare data, and they all come in different shapes and sizes. Each type has its own way of showing information, whether it is in a column, shape, or picture. To answer a question relating to a table, chart, or graph, some steps should be followed. First, the problem should be read thoroughly to determine what is being asked to determine what quantity is unknown. Then, the title of the table, chart, or graph should be read. The title should clarify what actual data is being summarized in the table. Next, look at the key and both the horizontal and vertical axis labels, if they are given. These items will provide information about how the data is organized. Finally, look to see if there is any more labeling inside the table. Taking the time to get a good idea of what the table is summarizing will be helpful as it is used to interpret information.

Tables are a good way of showing a lot of information in a small space. The information in a table is organized in columns and rows. For example, a table may be used to show the number of votes each candidate received in an election. By interpreting the table, one may observe which candidate won the election and which candidates came in second and third. In using a bar chart to display monthly rainfall amounts in different countries, rainfall can be compared between counties at different times of the year. Graphs are also a useful way to show change in variables over time, as in a line graph, or percentages of a whole, as in a pie graph.

The table below relates the number of items to the total cost. The table shows that 1 item costs $5. By looking at the table further, 5 items cost $25, 10 items cost $50, and 50 items cost $250. This cost can be extended for any number of items. Since 1 item costs $5, then 2 items would cost $10. Though this information isn't in the table, the given price can be used to calculate unknown information.

Number of Items	1	5	10	50
Cost ($)	5	25	50	250

A bar graph is a graph that summarizes data using bars of different heights. It is useful when comparing two or more items or when seeing how a quantity changes over time. It has both a horizontal and vertical axis. Interpreting **bar graphs** includes recognizing what each bar represents and connecting that to the two variables. The bar graph below shows the scores for six people on three different games. The color of the bar shows which game each person played, and the height of the bar indicates their score for that game. William scored 25 on game 3, and Abigail scored 38 on game 3. By comparing the bars, it's obvious that Williams scored lower than Abigail.

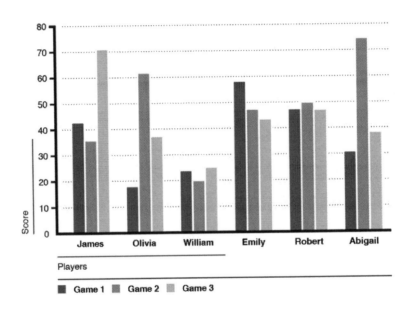

A line graph is a way to compare two variables. Each variable is plotted along an axis, and the graph contains both a horizontal and a vertical axis. On a **line graph,** the line indicates a continuous change. The change can be seen in how the line rises or falls, known as its slope, or rate of change. Often, in line graphs, the horizontal axis represents a variable of time. Audiences can quickly see if an amount has grown or decreased over time. The bottom of the graph, or the x-axis, shows the units for time, such as days, hours, months, etc. If there are multiple lines, a comparison can be made between what the two lines represent. For example, as shown previously, the following line graph shows the change in temperature over five days.

The top line represents the high, and the bottom line represents the low for each day. Looking at the top line alone, the high decreases for a day, then increases on Wednesday. Then it decreased on Thursday and increases again on Friday. The low temperatures have a similar trend, shown in bottom line. The range in temperatures each day can also be calculated by finding the difference between the top line and bottom line on a particular day. On Wednesday, the range was 14 degrees, from 62 to 76° F.

Daily Temperatures

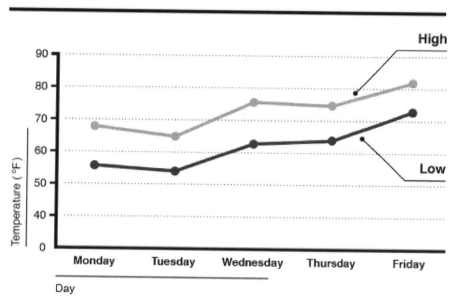

Pie charts are used to show percentages of a whole, as each category is given a piece of the pie, and together all the pieces make up a whole. They are a circular representation of data which are used to highlight numerical proportion. It is true that the arc length of each pie slice is proportional to the amount it individually represents. When a pie chart is shown, an audience can quickly make comparisons by comparing the sizes of the pieces of the pie. They can be useful for comparison between different categories. The following pie chart is a simple example of three different categories shown in comparison to each other.

Light gray represents cats, dark gray represents dogs, and the gray between those two represents other pets. As the pie is cut into three equal pieces, each value represents just more than 33 percent, or $\frac{1}{3}$ of the whole. Values 1 and 2 may be combined to represent $\frac{2}{3}$ of the whole. In an example where the total pie represents 75,000 animals, then cats would be equal to $\frac{1}{3}$ of the total, or 25,000. Dogs would equal 25,000 and other pets would also equal 25,000.

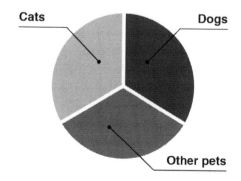

The fact that a circle is 360 degrees is used to create a pie chart. Because each piece of the pie is a percentage of a whole, that percentage is multiplied times 360 to get the number of degrees each piece represents. In the example above, each piece is 1/3 of the whole, so each piece is equivalent to 120 degrees. Together, all three pieces add up to 360 degrees.

Stacked bar graphs, also used fairly frequently, are used when comparing multiple variables at one time. They combine some elements of both pie charts and bar graphs, using the organization of bar graphs and the proportionality aspect of pie charts. The following is an example of a stacked bar graph that represents the number of students in a band playing drums, flute, trombone, and clarinet. Each bar graph is broken up further into girls and boys.

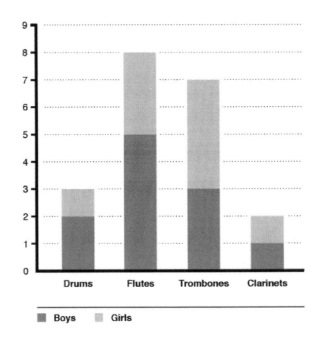

To determine how many boys play trombone, refer to the darker portion of the trombone bar, resulting in 3 students.

A **scatterplot** is another way to represent paired data. It uses Cartesian coordinates, like a line graph, meaning it has both a horizontal and vertical axis. Each data point is represented as a dot on the graph. The dots are never connected with a line. For example, the following is a scatterplot showing people's age versus height.

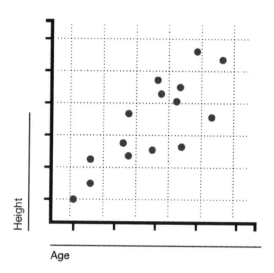

A scatterplot, also known as a **scattergram**, can be used to predict another value and to see if an association, known as a **correlation,** exists between a set of data. If the data resembles a straight line, the data is **associated.** The following is an example of a scatterplot in which the data does not seem to have an association:

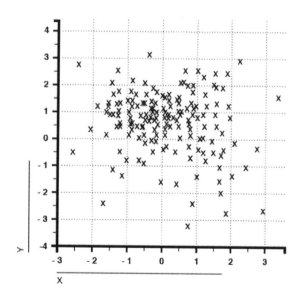

Sets of numbers and other similarly organized data can also be represented graphically. Venn diagrams are a common way to do so. A **Venn diagram** represents each set of data as a circle. The circles overlap, showing that each set of data is overlapping. A Venn diagram is also known as a **logic diagram** because it visualizes all possible logical combinations between two sets. Common elements of two sets are represented by the area of overlap. The following is an example of a Venn diagram of two sets A and B:

Parts of the Venn Diagram

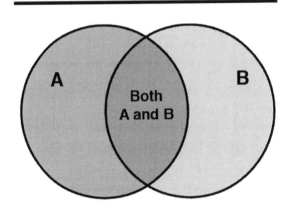

Another name for the area of overlap is the **intersection.** The intersection of A and B, $A \cap B$, contains all elements that are in both sets A and B. The **union** of A and B, $A \cup B$, contains all elements that are in either set A or set B. Finally, the **complement** of $A \cup B$ is equal to all elements that are not in either set A or set B. These elements are placed outside of the circles.

The following is an example of a Venn diagram in which 30 students were surveyed asking which type of siblings they had: brothers, sisters, or both. Ten students only had a brother, 7 students only had a sister, and 5 had both a brother and a sister. This number 5 is the intersection and is placed where the circles overlap. Two students did not have a brother or a sister. Two is therefore the complement and is placed outside of the circles.

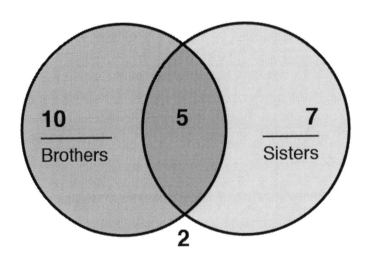

Venn diagrams can have more than two sets of data. The more circles, the more logical combinations are represented by the overlapping. The following is a Venn diagram that represents a different situation. Now, there were 30 students surveyed about the color of their socks. The innermost region represents those students that have green, pink, and blue socks on (perhaps a striped pattern). Therefore, 2 students had all three colors on their socks. In this example, all students had at least one of the three colors on their socks, so no one exists in the complement.

30 students

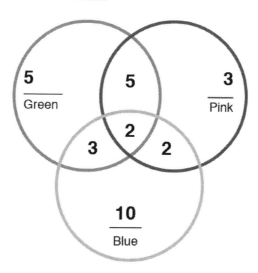

Venn diagrams are typically not drawn to scale, but if they are and their area is proportional to the amount of data it represents, it is known as an *area-proportional* Venn diagram.

Producing, Interpreting, Understanding, Evaluating, and Improving Models
Concrete models create a way of thinking about math that generates learning on a more permanent level. Memorizing abstract math formulas will not create a lasting effect on the brain. The following picture shows fractions represented by Lego blocks. By starting with the whole block of eight, it can be split into half, which is a four-block, and a fourth, which is a two-block. The one-eighth representation is a single block.

After splitting these up, addition and subtraction can be performed by adding or taking away parts of the blocks. Different combinations of fractions can be used to make a whole, or taken away to make various parts of a whole.

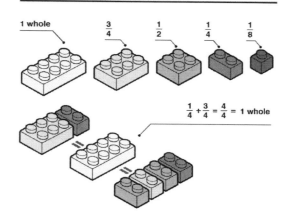

Using Colored Blocks to Model Functions

Multiplication can also be done using array models. The following picture shows a model of multiplying 3 times 4. **Arrays** are formed when the first factor is shown in a row. The second factor is shown in that number of columns. When the rectangle is formed, the blocks fill in to make a total, or the result of multiplication. The three rows and four columns show each factor and when the blocks are filled in, the total is twelve. Arrays are great ways to represent multiplication because they show each factor, and where the total comes from, with rows and columns.

Multiplication Array Model for 3 x 4

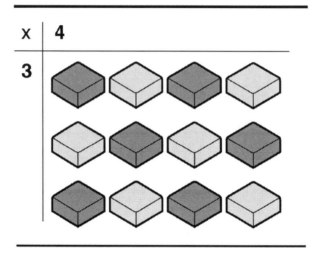

3 x 4 = 12

Another representation of fractions is shown below in the pie charts. Moving from whole numbers to part of numbers with fractions can be a concept that is difficult to grasp. Starting with a whole pie and splitting it into parts can be helpful with generating fractions. The first pie shows quarters or sections that are one-fourth because it is split into four parts. The second pie shows parts that equal one-fifth because it is split into five parts. Pies can also be used to add fractions. If $\frac{1}{5}$ and $\frac{1}{4}$ are being added, a common denominator must be found by splitting the pies into the same number of parts. The same number of parts can be found by determining the least common multiple. For 4 and 5, the least common multiple is 20. The pies can be split until there are 20 parts. The same portion of the pie can be shaded for each fraction and then added together to find the sum.

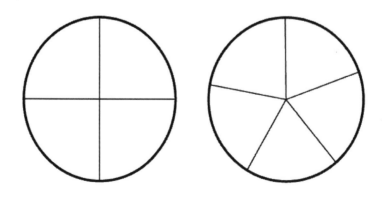

Illustrating and Explaining Multiplication and Division Problems Using Equations, Rectangular Arrays, and Area Models

Multiplication and division can be represented by equations. These equations show the numbers involved and the operation. For example, "eight multiplied by six equals forty-eight" is seen in the following equation: $8 \times 6 = 48$. This operation can be modeled by rectangular arrays where one factor, 8, is the number of rows, and the other factor, 6, is the number of columns, as follows:

Array of 8 x 6 = 48

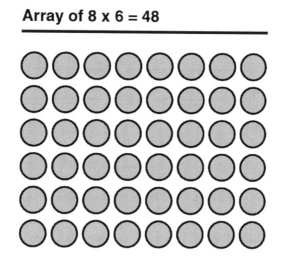

Rectangular arrays show what happens with the concept of multiplication. As one row of dots is drawn, that represents the first factor in the problem. Then the second factor is used to add the number of columns. The final model includes six rows of eight columns which results in forty-eight dots. These rectangular arrays show how multiplication of whole numbers will result in a number larger than the factors.

Division can also be represented by equations and area models. A division problem such as "twenty-four divided by three equals eight" can be written as the following equation:

$$24 \div 8 = 3$$

The object below shows an area model to represent the equation. As seen in the model, the whole box represents 24 and the 3 sections represent the division by 3. In more detail, there could be 24 dots written in the whole box and each box could have 8 dots in it.

Division shows how numbers can be divided into groups. For the example problem, it is asking how many numbers will be in each of the 3 groups that together make 24. The answer is 8 in each group.

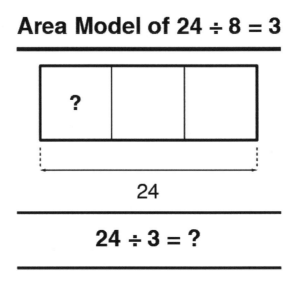

Area Model of 24 ÷ 8 = 3

24

24 ÷ 3 = ?

Converting Within and Between Standard and Metric Systems

When working with dimensions, sometimes the given units don't match the formula, and conversions must be made. The metric system has base units of meter for length, kilogram for mass, and liter for liquid volume. This system expands to three places above the base unit and three places below. These places correspond with prefixes with a base of 10.

The following table shows the conversions:

kilo-	hecto-	deca-	base	deci-	centi-	milli-
1,000 times the base	100 times the base	10 times the base		1/10 times the base	1/100 times the base	1/1000 times the base

To convert between units within the metric system, values with a base ten can be multiplied. The decimal can also be moved in the direction of the new unit by the same number of zeros on the number. For example, 3 meters is equivalent to .003 kilometers. The decimal moved three places (the same number of zeros for kilo-) to the left (the same direction from base to kilo-). Three meters is also equivalent to 3,000 millimeters. The decimal is moved three places to the right because the prefix milli- is three places to the right of the base unit.

The English Standard system used in the United States has a base unit of foot for length, pound for weight, and gallon for liquid volume. These conversions aren't as easy as the metric system because they aren't a base ten model. The following table shows the conversions within this system.

Length	Weight	Capacity
1 foot (ft) = 12 inches (in) 1 yard (yd) = 3 feet 1 mile (mi) = 5280 feet 1 mile = 1760 yards	1 pound (lb) = 16 ounces (oz) 1 ton = 2000 pounds	1 tablespoon (tbsp) = 3 teaspoons (tsp) 1 cup (c) = 16 tablespoons 1 cup = 8 fluid ounces (oz) 1 pint (pt) = 2 cups 1 quart (qt) = 2 pints 1 gallon (gal) = 4 quarts

When converting within the English Standard system, most calculations include a conversion to the base unit and then another to the desired unit. For example, take the following problem: $3 \ quarts = $ ___ $cups$. There is no straight conversion from quarts to cups, so the first conversion is from quarts to pints. There are 2 pints in 1 quart, so there are 6 pints in 3 quarts. This conversion can be solved as a proportion: $\frac{3 \ qt}{x} = \frac{1 \ qt}{2 \ pints}$. It can also be observed as a ratio 2:1, expanded to 6:3. Then the 6 pints must be converted to cups. The ratio of pints to cups is 1:2, so the expanded ratio is 6:12. For 6 pints, the measurement is 12 cups. This problem can also be set up as one set of fractions to cancel out units. It begins with the given information and cancels out matching units on top and bottom to yield the answer. Consider the following expression:

$$\frac{3 \ quarts}{1} \times \frac{2 \ pints}{1 \ quart} \times \frac{2 \ cups}{1 \ pint}$$

It's set up so that units on the top and bottom cancel each other out:

$$\frac{3 \ \cancel{quarts}}{1} \times \frac{2 \ \cancel{pints}}{1 \ \cancel{quart}} \times \frac{2 \ cups}{1 \ \cancel{pint}}$$

The numbers can be calculated as $3 \times 3 \times 2$ on the top and 1 on the bottom. It still yields an answer of 12 cups.

This process of setting up fractions and canceling out matching units can be used to convert between standard and metric systems. A few common equivalent conversions are 2.54 cm = 1 inch, 3.28 feet = 1 meter, and 2.205 pounds = 1 kilogram. Writing these as fractions allows them to be used in conversions. For the fill-in-the-blank problem 5 meters = ___ feet, an expression using conversions starts with the expression $\frac{5 \ meters}{1} \times \frac{3.28 \ feet}{1 \ meter}$, where the units of meters will cancel each other out, and the final unit is feet. Calculating the numbers yields 16.4 feet. This problem only required two fractions. Others may require longer expressions, but the underlying rule stays the same. When there's a unit on the top of the fraction that's the same as the unit on the bottom, then they cancel each other out. Using this logic and the conversions given above, many units can be converted between and within the different systems.

The conversion between Fahrenheit and Celsius is found in a formula:

$$°C = (°F - 32) \times \frac{5}{9}$$

For example, to convert 78°F to Celsius, the given temperature would be entered into the formula:

$$°C = (78 - 32) \times \frac{5}{9}$$

Solving the equation, the temperature comes out to be 25.56°C. To convert in the other direction, the formula becomes:

$$°F = °C * \frac{9}{5} + 32$$

Remember the order of operations when calculating these conversions.

Solving Problems Involving Elapsed Time, Money, Length, Volume, and Mass

To solve problems, follow these steps: Identify the variables that are known, decide which equation should be used, substitute the numbers, and solve. To solve an equation for the amount of time that has elapsed since an event, use the equation T = L – E where T represents the elapsed time, L represents the later time, and E represents the earlier time. For example, the Minnesota Vikings have not appeared in the Super Bowl since 1976. If the year is now 2017, how long has it been since the Vikings were in the Super Bowl? The later time, L, is 2017, E = 1976 and the unknown is T.

Substituting these numbers, the equation is T = 2017 – 1976, and so T = 41. It has been 41 years since the Vikings have appeared in the Super Bowl. Questions involving total cost can be solved using the formula, C = I + T where C represents the total cost, I represents the cost of the item purchased, and T represents the tax amount. To find the length of a rectangle given the area = 32 square inches and width = 8 inches, the formula A = L × W can be used. Substitute 32 for A and substitute 8 for w, giving the equation 32 = L×8. This equation is solved by dividing both sides by 8 to find that the length of the rectangle is 4. The formula for volume of a rectangular prism is given by the equation:

$$V = L \times W \times H$$

If the length of a rectangular juice box is 4 centimeters, the width is 2 centimeters, and the height is 8 centimeters, what is the volume of this box? Substituting in the formula we find V = 4 × 2 × 8, so the volume is 64 cubic centimeters. In a similar fashion as those previously shown, the mass of an object can be calculated given the formula, Mass = Density × Volume.

Measuring and Comparing Lengths of Objects Using Standard Tools

Lengths of objects can be measured using tools such as rulers, yard sticks, meter sticks, and tape measures. Typically, a ruler measures 12 inches, or one foot. For this reason, a ruler is normally used to measure lengths smaller than or just slightly more than 12 inches. Rulers may represent centimeters instead of inches. Some rulers have inches on one side and centimeters on the other. Be sure to recognize what units you are measuring in. The standard ruler measurements are divided into units of 1 inch and normally broken down to $\frac{1}{2}, \frac{1}{4}, \frac{1}{8}$, and even $\frac{1}{16}$ of an inch for more precise measurements. If measuring in centimeters, the centimeter is likely divided into tenths.

To measure the size of a picture, for purposes of buying a frame, a ruler is helpful. If the picture is very large, a yardstick, which measures 3 feet and normally is divided into feet and inches, might be useful. Using metric units, the meter stick measures 1 meter and is divided into 100 centimeters. To measure the size of a window in a home, either a yardstick or meter stick would work. To measure the size of a room, though, a tape measure would be the easiest tool to use. Tape measures can measure up to 10 feet, 25 feet, or more depending on the particular tape measure.

Comparing Relative Sizes of U.S. Customary Units and Metric Units

Measuring length in United States customary units is typically done using inches, feet, yards, and miles. When converting among these units, remember that 12 inches = 1 foot, 3 feet = 1 yard, and 5280 feet = 1 mile. Common customary units of weight are ounces and pounds. The conversion needed is 16 ounces = 1 pound. For customary units of volume ounces, cups, pints, quarts, and gallons are typically used. For conversions, use 8 ounces = 1 cup, 2 cups = 1 pint, 2 pints = 1 quart, and 4 quarts = 1 gallon. For measuring lengths in metric units, know that 100 centimeters = 1 meter, and 1000 meters = 1 kilometer. For metric units of measuring weights, grams and kilograms are often used. Know that 1000 grams = 1 kilogram when making conversions. For metric measures of volume, the most common units are milliliters and liters. Remember that 1000 milliliters = 1 liters.

Practice Test #1

1. What is $\frac{12}{60}$ converted to a percentage?
   a. 0.20
   b. 20%
   c. 25%
   d. 12%
   e. 1.2%

2. Which of the following is the correct decimal form of the fraction $\frac{14}{33}$ rounded to the nearest hundredth place?
   a. 0.420
   b. 0.14
   c. 0.424
   d. 0.140
   e. 0.42

3. Which of the following represents the correct sum of $\frac{14}{15}$ and $\frac{2}{5}$?
   a. $\frac{20}{15}$
   b. $\frac{4}{3}$
   c. $\frac{16}{20}$
   d. $\frac{4}{5}$
   e. $\frac{16}{15}$

4. What is the product of $\frac{5}{14}$ and $\frac{7}{20}$?
   a. $\frac{1}{8}$
   b. $\frac{35}{280}$
   c. $\frac{12}{34}$
   d. $\frac{1}{2}$
   e. $\frac{7}{140}$

5. Which of the following is incorrect?

a. $-\frac{1}{5} < \frac{4}{5}$

b. $\frac{4}{5} > -\frac{1}{5}$

c. $-\frac{1}{5} > \frac{4}{5}$

d. $\frac{1}{5} > -\frac{4}{5}$

e. $\frac{4}{5} > \frac{1}{5}$

6. What is the solution to the equation $3(x + 2) = 14x - 5$?

a. $x = 1$

b. $x = -1$

c. $x = 0$

d. All real numbers

e. No solution

7. Which of the following is the result when solving the equation $4(x + 5) + 6 = 2(2x + 3)$?

a. $x = 6$

b. $x = 1$

c. $x = 26$

d. All real numbers

e. No solution

8. How many cases of cola can Lexi purchase if each case is $3.50 and she has $40?

a. 10

b. 12

c. 11.4

d. 11

e. 12.5

9. Two consecutive integers exist such that the sum of three times the first and two less than the second is equal to 411. What are those integers?

a. 103 and 104

b. 104 and 105

c. 102 and 103

d. 100 and 101

e. 101 and 102

10. In a neighborhood, 15 out of 80 of the households have children under the age of 18. What percentage of the households have children?

a. 0.1875%

b. 18.75%

c. 1.875%

d. 15%

e. 1.50%

11. Paul took a written driving test, and he got 12 of the questions correct. If he answered 75% of the total questions correctly, how many problems were there in the test?

    a. 25
    b. 15
    c. 20
    d. 18
    e. 16

12. If a car is purchased for $15,395 with a 7.25% sales tax, how much is the total price?
    a. $15,395.07
    b. $16,511.14
    c. $16,411.13
    d. $15,402
    e. $16,113.10

13. Which equation correctly shows how to find the surface area of a cylinder?

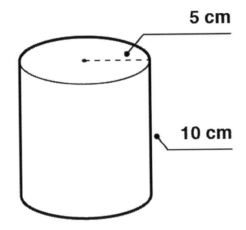

5 cm

10 cm

    a. $SA = 2\pi \times 5 \times 10 + 2(\pi 5^2)$
    b. $SA = 5 \times 2\pi * 5$
    c. $SA = 2\pi 5^2$
    d. $SA = 2\pi \times 10 + \pi 5^2$
    e. $SA = 2\pi \times 5 \times 10 + \pi 5^2$

14. Which shapes could NOT be used to compose a hexagon?

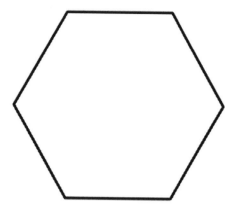

    a. Six triangles
    b. One rectangle and two triangles
    c. Two rectangles
    d. Two trapezoids
    e. One rectangle and four triangles

15. A grocery store sold 48 bags of apples in one day, and 9 of the bags contained Granny Smith apples. The rest contained Red Delicious apples. What is the ratio of bags of Granny Smith to bags of Red Delicious that were sold?
    a. 48:9
    b. 39:9
    c. 9:48
    d. 9:39
    e. 39:48

16. If Oscar's bank account totaled $4,000 in March and $4,900 in June, what was the rate of change in his bank account total over those three months?
    a. $900 a month
    b. $300 a month
    c. $4,900 a month
    d. $100 a month
    e. $4,000 a month

17. Erin and Katie work at the same ice cream shop. Together, they always work less than 21 hours a week. In a week, if Katie worked two times as many hours as Erin, how many hours could Erin work?
    a. Less than 7 hours
    b. Less than or equal to 7 hours
    c. More than 7 hours
    d. Less than 8 hours
    e. More than 8 hours

18. From the chart below, which two are preferred by more men than women?

Preferred Movie Genres

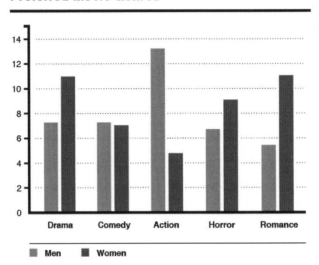

a. Comedy and Action
b. Drama and Comedy
c. Action and Horror
d. Action and Romance
e. Romance and Comedy

19. Which type of graph best represents a continuous change over a period of time?
a. Stacked bar graph
b. Bar graph
c. Pie graph
d. Histogram
e. Line graph

20. What is the mode for the grades shown in the chart below?

Science Grades	
Jerry	65
Bill	95
Anna	80
Beth	95
Sara	85
Ben	72
Jordan	98

a. 65
b. 33
c. 95
d. 90
e. 84.3

21. What is the area of the shaded region?

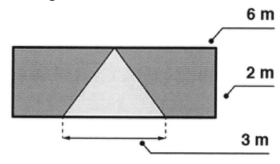

6 m

2 m

3 m

    a. 9 m²
    b. 12 m²
    c. 6 m²
    d. 8 m²
    e. 4.5 m²

22. What is the volume of the cylinder below?

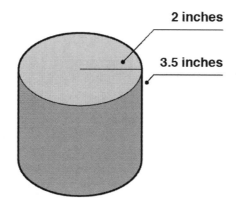

2 inches

3.5 inches

    a. 18.84 in³
    b. 45.00 in³
    c. 70.43 in³
    d. 43.96 in³
    e. 21.98 in³

23. How many centimeters are in 3 feet? (Note: 2.54cm = 1 inch)
    a. 0.635
    b. 1.1811
    c. 14.17
    d. 7.62
    e. 91.44

24. Which of the following relations is a function?
    a. {(1, 4), (1, 3), (2, 4), (5, 6)}
    b. {(-1, -1), (-2, -2), (-3, -3), (-4, -4)}
    c. {(0, 0), (1, 0), (2, 0), (1, 1)}
    d. {(1, 0), (1, 2), (1, 3), (1, 4)}
    e. {(-1, 1), (1, -3), (2, 7), (-1, 6)}

25. Find the indicated function value: $f(5)$ for $f(x) = x^2 - 2x + 1$.
    a. 16
    b. 1
    c. 5
    d. 8
    e. Does not exist

26. What is the domain of $f(x) = 4x^2 + 2x - 1$?
    a. $(0, \infty)$
    b. $(-\infty, 0)$
    c. $(-\infty, 4)$
    d. $(-1, 4)$
    e. $(-\infty, \infty)$

27. The function $f(x) = 3.1x + 240$ models the total U.S. population, in millions, x years after the year 1980. Use this function to answer the following question: What is the total U.S. population in 2011? Round to the nearest million.
    a. 336 people
    b. 336 million people
    c. 6,474 people
    d. 647 million people
    e. 64 million people

28. What is the label given to a problem that multiplies a matrix by a constant?
    a. Vector multiplication
    b. Scalar multiplication
    c. Inverse of a matrix
    d. Transposition of a matrix
    e. Product of a matrix

29. What is the range of the polynomial function $f(x) = 2x^2 + 5$?
    a. $(-\infty, \infty)$
    b. $(2, \infty)$
    c. $(0, \infty)$
    d. $(-\infty, 5)$
    e. $[5, \infty)$

30. What are the two values that always describe a vector?
    a. Magnitude and direction
    b. Magnitude and length
    c. Length and position
    d. Direction and position
    e. Magnitude and position

31. For which two values of x is $g(x) = 4x + 4$ equal to $g(x) = x^2 + 3x + 2$?
    a. 1, 0
    b. 2, -1
    c. -1, 2
    d. 1, 2
    e. -2, 1

32. The population of coyotes in the local national forest has been declining since 2000. The population can be modeled by the function $y = -(x - 2)^2 + 1600$, where y represents number of coyotes and x represents the number of years past 2000. When will there be no more coyotes?
   a. 2020
   b. 2040
   c. 2012
   d. 2064
   e. 2042

33. A ball is thrown up from a building that is 800 feet high. Its position s in feet above the ground is given by the function $s = -32t^2 + 90t + 800$, where t is the number of seconds since the ball was thrown. How long will it take for the ball to come back to its starting point? Round your answer to the nearest tenth of a second.
   a. 0 seconds
   b. 2.8 seconds
   c. 3 seconds
   d. 8 seconds
   e. 1.5 seconds

34. What are the zeros of the following cubic function?
$$g(x) = x^3 - 2x^2 - 9x + 18$$
   a. 2, 3
   b. 2, 3, -2
   c. 2, 3, -3,
   d. 2, -2
   e. 0, 2, 3

35. What is the domain of the following rational function?
$$f(x) = \frac{x^3 + 2x + 1}{2 - x}$$
   a. $(-\infty, -2) \cup (-2, \infty)$
   b. $(-\infty, 2) \cup (2, \infty)$
   c. $(2, \infty)$
   d. $(-2, \infty)$
   e. $(-2, 2)$

36. Given the function $f(x) = 4x - 2$, what is the correct form of the simplified difference quotient:
$$\frac{f(x + h) - f(x)}{h}$$
   a. $4x - 1$
   b. $4x$
   c. 4
   d. $4x + h$
   e. $2x - 1$

37. Which set of matrices represents the following system of equations?

$$\begin{cases} x - 2y + 3z = 7 \\ 2x + y + z = 4 \\ -3x + 2y - 2z = -10 \end{cases}$$

a. $\begin{bmatrix} 1 & -2 & 3 \\ 2 & 3 & 1 \\ -3 & 1 & 2 \end{bmatrix}\begin{bmatrix} x \\ y \\ z \end{bmatrix} = \begin{bmatrix} 7 \\ -4 \\ 10 \end{bmatrix}$

b. $\begin{bmatrix} 1 & 2 & -3 \\ -2 & 1 & 2 \\ 3 & 1 & -2 \end{bmatrix}\begin{bmatrix} x \\ y \\ z \end{bmatrix} = \begin{bmatrix} 7 \\ 4 \\ -10 \end{bmatrix}$

c. $\begin{bmatrix} 1 & -2 & 4 \\ -3 & 1 & 7 \\ 2 & 2 & -10 \end{bmatrix}\begin{bmatrix} 3 \\ 1 \\ -2 \end{bmatrix} = \begin{bmatrix} x \\ y \\ z \end{bmatrix}$

d. $\begin{bmatrix} 1 & -2 & 3 \\ 2 & 1 & 1 \\ -3 & 2 & -2 \end{bmatrix}\begin{bmatrix} x \\ y \\ z \end{bmatrix} = \begin{bmatrix} 7 \\ 4 \\ -10 \end{bmatrix}$

e. $\begin{bmatrix} -1 & 2 & -3 \\ 2 & 1 & 1 \\ -3 & 2 & 2 \end{bmatrix}\begin{bmatrix} x \\ y \\ z \end{bmatrix} = \begin{bmatrix} 5 \\ 4 \\ -8 \end{bmatrix}$

38. Which expression is equivalent to $\sqrt[4]{x^6} - \frac{x}{x^3} + x - 2$?

a. $x^{\frac{3}{2}} - x^2 + x - 2$

b. $x^{\frac{2}{3}} - x^{-2} + x - 2$

c. $x^{\frac{3}{2}} - \frac{1}{x^2} + x - 2$

d. $x^{\frac{2}{3}} - \frac{1}{x^2} + x - 2$

e. $x^{\frac{1}{3}} - \frac{1}{x^2} + x - 2$

39. Which of the following is perpendicular to the line $4x + 7y = 23$?

a. $y = -\frac{4}{7}x + 23$

b. $y = \frac{7}{4}x - 12$

c. $4x + 7y = 14$

d. $y = -\frac{7}{4}x + 11$

e. $y = \frac{4}{7}x - 12$

40. What is the solution to the following system of equations?
$$2x - y = 6$$
$$y = 8x$$

    a. (1, 8)
    b. (-1, -8)
    c. (-1, 8)
    d. All real numbers.
    e. There is no solution.

41. The following set represents the test scores from a university class: {35, 79, 80, 87, 87, 90, 92, 95, 95, 98, 99}. If the outlier is removed from this set, which of the following is TRUE?
    a. The mean and the median will decrease.
    b. The mean and the median will increase.
    c. The mean and the mode will increase.
    d. The mean and the mode will decrease.
    e. The mean, median, and mode will increase.

42. The mass of the moon is about 7.348×10^{22} kilograms and the mass of Earth is 5.972×10^{24} kilograms. How many times GREATER is Earth's mass than the moon's mass?
    a. 8.127×10^1
    b. 8.127
    c. 812.7
    d. 8.127×10^{-1}
    e. 0.8127

43. Which of the following displays the function that the following graph represents?

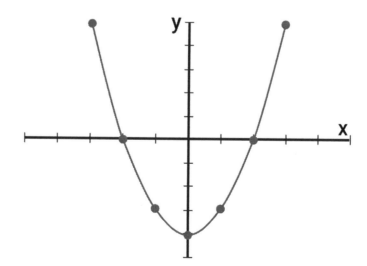

    a. $f(x) = 2x^2 - 4$
    b. $f(x) = 4x^2 - 4$
    c. $f(x) = x^2 + 4$
    d. $f(x) = x^2 - 4$
    e. $f(x) = 2x^2 - 4$

44. The points (1, 3) and (-2, -5) lie in the Cartesian coordinate plane. What are the coordinates of the midpoint of the line segment that connects these two points?

    a. (-1, -1)

    b. (-1, -2)

    c. $\left(-\frac{1}{2}, -1\right)$

    d. $\left(-\frac{1}{2}, -2\right)$

    e. $\left(-1, -\frac{1}{2}\right)$

45. A trapezoid comprises two parallel bases of length 4 cm and 6 cm, respectively. Also, its height is 9 cm. What is the area of this trapezoid?

    a. 90 square cm

    b. 45 square cm

    c. 24 square cm

    d. 180 square cm

    e. 54 square cm

46. In the cartesian coordinate plane, what is the diameter of the circle with the origin as the center and (4, 5) as one of the endpoints of the radius?

    a. 41

    b. $\sqrt{41}$

    c. $\sqrt{45}$

    d. $3\sqrt{41}$

    e. $2\sqrt{41}$

47. A rectangle has a diagonal with length 16 cm. If the width of the rectangle is 8 cm, what is the area of the rectangle?

    a. 64 square cm

    b. 192 square cm

    c. $64\sqrt{3}$ square cm

    d. $18\sqrt{3}$ square cm

    e. 18 square cm

48. The three interior angles of a triangle measure $3x$, $4x$, and $8x$ degrees. What must the measure of the three angles be?

    a. 24, 36, and 48 degrees

    b. 45, 45, and 90 degrees

    c. 36, 48, and 76 degrees

    d. 36, 48, and 96 degrees

    e. 40, 70, and 80 degrees

49. What is the simplified form of $(4y^3)^4(3y^7)^2$?

    a. $12y^{26}$

    b. $2304y^{16}$

    c. $12y^{14}$

    d. $2304y^{26}$

    e. $12y^{16}$

50. Use the graph below entitled "Projected Temperatures for Tomorrow's Winter Storm" to answer the question.

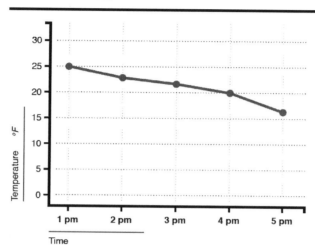

Projected Temperatures for Tomorrow's Winter Storm

What is the expected temperature at 3:00 p.m.?
   a. 25 degrees
   b. 22 degrees
   c. 20 degrees
   d. 16 degrees
   e. 18 degrees

51. The number of members of the House of Representatives varies directly with the total population in a state. If the state of New York has 19,800,000 residents and has 27 total representatives, how many should Ohio have with a population of 11,800,000?
   a. 10
   b. 9
   c. 11
   d. 5
   e. 12

52. Eva Jane is practicing for an upcoming 5K run. She has recorded the following times (in minutes):
      25, 18, 23, 28, 30, 22.5, 23, 33, 20
Use the above information to answer the next three questions to the closest minute. What is Eva Jane's mean time?
   a. 26 minutes
   b. 19 minutes
   c. 24.5 minutes
   d. 23 minutes
   e. 25 minutes

53. What is the mode of Eva Jane's time?
    a. 16 minutes
    b. 20 minutes
    c. 23 minutes
    d. 33 minutes
    e. 25 minutes

54. What is Eva Jane's median score?
    a. 23 minutes
    b. 17 minutes
    c. 28 minutes
    d. 19 minutes
    e. 25 minutes

55. What is the perimeter of the following figure?

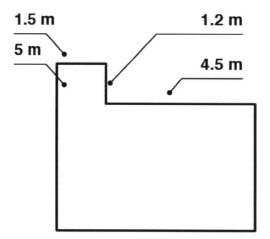

    a. 13.4 m
    b. 22 m
    c. 12.2 m
    d. 22.5 m
    e. 24.4 m

56. Given that $abc \neq 0$, which of the following is equivalent to the expression $\frac{a^5 b^3 c^7}{(ab^2 c^3)^2}$?
    a. $a^3 bc$

    b. $a^3 b^2 c$

    c. $\frac{a^3 c}{b}$

    d. $\frac{ac}{b}$

    e. $\frac{ab^3}{c}$

57. A shoe salesclerk earns $8x + 0.00048y^2$ dollars, where x is the total number of hours worked and y represents her total sales in dollars. If she earned \$200 dollars and had \$500 in sales, how many hours did she work?

    a. 8
    b. 10
    c. 12
    d. 14
    e. 16

58. If $x < 0$ and $6x^2 - 14x - 20 = 0$, then what is x?

    a. -1
    b. $-\frac{10}{3}$
    c. -20
    d. $-\frac{1}{3}$
    e. -3

59. A system of linear equations in two variables has no solutions. One of the lines is drawn below. Which of the following equations could the other line be?

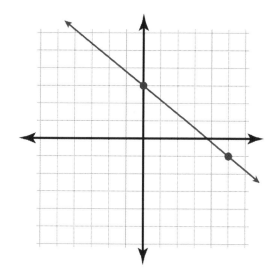

    a. $y = \frac{5}{4}x + 4$
    b. $y = -\frac{4}{5}x + 16$
    c. $y = 5x + 3$
    d. $y = -\frac{5}{4}x + 3$
    e. $y = -4x + 2$

60. What are three consecutive odd integers that satisfy the following: the sum of four times the first, two times the second, and two less than the third is equal to 111?

    a. 9, 11, 13
    b. 11, 13, 15
    c. 13, 15, 19
    d. 15, 17, 19
    e. 21, 23, 24

61. In a graduate school classroom, the average age of the male students is 27 and the average age of the female members is 24. If 30% of the members are male, what is the average age of the entire class?
    a. 25.5
    b. 26
    c. 24
    d. 24.9
    e. 22.1

62. In a small university, the ratio of male to female students is 3:2. If there are 900 total students, how many are men?
    a. 540
    b. 360
    c. 750
    d. 150
    e. 600

63. A bag contains some colored soccer balls: 9 red balls, 8 orange balls, and 5 green balls. If a ball is randomly selected from the bag, what is the probability that an orange ball is NOT selected?
    a. $\frac{8}{23}$
    b. $\frac{8}{17}$
    c. $\frac{15}{23}$
    d. $\frac{5}{23}$
    e. $\frac{5}{13}$

64. The probability of randomly selecting a red card from a bag of cards is $\frac{7}{10}$. If there are 210 cards in the bag, how many are NOT red?
    a. 147
    b. 63
    c. 210
    d. 70
    e. 130

Answer Explanations #1

1. B: The fraction $\frac{12}{60}$ can be reduced to $\frac{1}{5}$ in lowest terms. First, it must be converted to a decimal. Dividing 1 by 5 results in 0.2. Then, to convert to a percentage, move the decimal point two units to the right and add the percentage symbol. The result is 20%.

2. E: If a calculator is used, divide 33 into 14 and keep two decimal places. If a calculator is not used, multiply both the numerator and denominator times 3. This results in the fraction $\frac{42}{99}$, and hence a decimal of 0.42.

3. B: Common denominators must be used. The LCD is 15, and $\frac{2}{5} = \frac{6}{15}$. Therefore, $\frac{14}{15} + \frac{6}{15} = \frac{20}{15}$, and in lowest terms the answer is $\frac{4}{3}$. A common factor of 5 was divided out of both the numerator and denominator.

4. A: A product is found by multiplication. Multiplying two fractions together is easier when common factors are cancelled first to avoid working with larger numbers.

$$\frac{5}{14} \times \frac{7}{20} = \frac{5}{2 \times 7} \times \frac{7}{5 \times 4}$$

$$\frac{1}{2} \times \frac{1}{4} = \frac{1}{8}$$

5. C: $-\frac{1}{5} > \frac{4}{5}$ is incorrect. The expression on the left is negative, which means that it is smaller than the expression on the right. As it is written, the inequality states that the expression on the left is greater than the expression on the right, which is not true.

6. A: First, the distributive property must be used on the left side. This results in:

$$3x + 6 = 14x - 5$$

The addition property is then used to add 5 to both sides, and then to subtract $3x$ from both sides, resulting in $11 = 11x$. Finally, the multiplication property is used to divide each side by 11. Therefore, $x = 1$ is the solution.

7. E: The distributive property is used on both sides to obtain:

$$4x + 20 + 6 = 4x + 6$$

Then, like terms are collected on the left, resulting in"

$$4x + 26 = 4x + 6$$

Next, the addition principle is used to subtract $4x$ from both sides, and this results in the false statement $26 = 6$. Therefore, there is no solution.

8. D: This is a one-step real-world application problem. The unknown quantity is the number of cases of cola to be purchased. Let x be equal to this amount. Because each case costs $3.50, the total number of cases times $3.50 must equal $40. This translates to the mathematical equation $3.5x = 40$. Divide both sides by 3.5 to obtain $x = 11.4286$, which has been rounded to four decimal places. Because cases are sold whole (the store does not sell portions of cases), and there is not enough money to purchase 12 cases, there is only enough money to purchase 11.

9. A: First, the variables have to be defined. Let x be the first integer and therefore $x + 1$ is the second integer. This is a two-step problem. The sum of three times the first and two less than the second is translated into the following expression:

$$3x + (x + 1 - 2)$$

This expression is set equal to 411 to obtain:

$$3x + (x + 1 - 2) = 412$$

The left-hand side is simplified to obtain:

$$4x - 1 = 411$$

The addition and multiplication properties are used to solve for x. First, add 1 to both sides and then divide both sides by 4 to obtain $x = 103$. The next consecutive integer is 104.

10. B: First, the information is translated into the ratio $\frac{15}{80}$. To find the percentage, translate this fraction into a decimal by dividing 15 by 80. The corresponding decimal is 0.1875. Move the decimal point two units to the right to obtain the percentage 18.75%.

11. E: The unknown quantity is the number of total questions on the test. Let x be equal to this unknown quantity. Therefore:

$$0.75x = 12$$

Divide both sides by 0.75 to obtain $x = 16$.

12. B: If sales tax is 7.25%, the price of the car must be multiplied times 1.0725 to account for the additional sales tax. Therefore:

$$15{,}395 \times 1.0725 = 16{,}511.1375$$

This amount is rounded to the nearest cent, which is:

$$\$16{,}511.14$$

13. A: The surface area for a cylinder is the sum of the two circle bases and the rectangle formed on the side. This is easily seen in the net of a cylinder.

The Net of a Cylinder

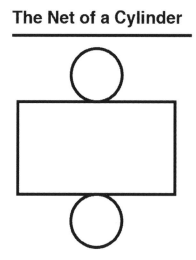

The area of a circle is found by multiplying pi times the radius squared. The rectangle's area is found by multiplying the circumference of the circle by the height. The equation $SA = 2\pi \times 5 \times 10 + 2(\pi 5^2)$ shows the area of the rectangle as $2\pi \times 5 \times 10$, which yields 314. The area of the bases is found by $\pi 5^2$, which yields 78.5, then multiplied by 2 for the two bases.

14. C: A hexagon can be formed by any combination of the given shapes except for two rectangles. There are no two rectangles that can make up a hexagon.

15. D: There were 48 total bags of apples sold. If 9 bags were Granny Smith and the rest were Red Delicious, then 48 − 9 = 39 bags were Red Delicious. Therefore, the ratio of Granny Smith to Red Delicious is 9:39.

16. B: The average rate of change is found by calculating the difference in dollars over the elapsed time. Therefore, the rate of change is equal to $4,900−$4,000÷3 months, which is equal to $900÷3 or $300 a month.

17. A: Let x be the unknown, the number of hours Erin can work. We know Katie works $2x$, and the sum of all hours is less than 21. Therefore, $x + 2x < 21$, which simplifies into $3x < 21$. Solving this results in the inequality $x < 7$ after dividing both sides by 3. Therefore, Erin can work less than 7 hours.

18. A: The chart is a bar chart showing how many men and women prefer each genre of movies. The dark gray bars represent the number of women, while the light gray bars represent the number of men. The light gray bars are higher and represent more men than women for the genres of Comedy and Action.

19. E: A line graph represents continuous change over time. The line on the graph is continuous and not broken, as on a scatter plot. Stacked bar graphs are used when comparing multiple variables at one time. They combine some elements of both pie charts and bar graphs, using the organization of bar graphs and the proportionality aspect of pie charts. A bar graph may show change but isn't necessarily continuous over time. A pie graph is better for representing percentages of a whole. Histograms are best used in grouping sets of data in bins to show the frequency of a certain variable.

20. C: The mode for a set of data is the value that occurs the most. The grade that appears the most is 95. It's the only value that repeats in the set. The mean is around 84.3.

21. A: The area of the shaded region is calculated in a few steps. First, the area of the rectangle is found using the formula:

$$A = length \times width = 6 \times 2 = 12$$

Second, the area of the triangle is found using the formula:

$$A = \frac{1}{2} \times base \times height$$

$$\frac{1}{2} \times 3 \times 2 = 3$$

The last step is to take the rectangle area and subtract the triangle area. The area of the shaded region is:

$$A = 12 - 3 = 9m^2$$

22. D: The volume for a cylinder is found by using the formula:

$$V = \pi r^2 h$$

$$\pi(2^2) \times 3.5 = 43.96 in^3$$

23. E: The conversion between feet and centimeters requires a middle term. As there are 2.54 centimeters in 1 inch, the conversion between inches and feet must be found. As there are 12 inches in a foot, the fractions can be set up as follows:

$$3\, feet \times \frac{12\, inches}{1\, foot} \times \frac{2.54\, cm}{1\, inch}$$

The feet and inches cancel out to leave only centimeters for the answer. The numbers are calculated across the top and bottom to yield:

$$\frac{3 \times 12 \times 2.54}{1 \times 1} = 91.44$$

The number and units used together form the answer of 91.44 cm.

24. B: The only relation in which every *x*-value corresponds to exactly one *y*-value is the relation given in *B*, making it a function. The other relations have the same first component paired up to different second components, which goes against the definition of functions.

25. A: To find a function value, plug in the number given for the variable and evaluate the expression, using the order of operations (parentheses, exponents, multiplication, division, addition, subtraction). The function given is a polynomial function and:

$$f(5) = 5^2 - 2 \times 5 + 1$$

$$25 - 10 + 1 = 16$$

26. E: The function given is a polynomial function. Anything can be plugged into a polynomial function to get an output. Therefore, its domain is all real numbers, which is expressed in interval notation as $(-\infty, \infty)$.

27. B: The variable x represents the number of years after 1980. The year 2011 was 31 years after 1980, so plug 31 into the function to obtain:

$$f(31) = 3.1 \times 31 + 240 = 336.1$$

This value rounds to 336, and represents 336 million people.

28. B: The correct answer is Choice *B* because multiplying a matrix by a constant is called scalar multiplication. A scalar is a constant number, which means the only thing it changes about a matrix is its magnitude. For a given matrix, $\begin{bmatrix} 3 & 4 \\ 6 & 5 \end{bmatrix}$, scalar multiplication can be applied by multiplying by 2, which yields the matrix $\begin{bmatrix} 6 & 8 \\ 12 & 10 \end{bmatrix}$. Notice that the dimensions of the matrix did not change, just the magnitude of the numbers.

29. E: This is a parabola that opens up, as the coefficient on the x^2 term is positive. The smallest number in its range occurs when plugging 0 into the function $f(0) = 5$. Any other output is a number larger than 5, even when a positive number is plugged in. When a negative number gets plugged into the function, the output is positive, and same with a positive number. Therefore, the domain is written as $[5, \infty)$ in interval notation.

30. A: The vector is described as having both magnitude and direction. The magnitude is the size of the vector and the direction is the path along with which the force is being applied. The second answer choice has magnitude and length, which are essentially the same. The third and fourth answer choices include length and position, but position is not part of the description of a vector.

31. C: First set the functions equal to one another, resulting in:

$$x^2 + 3x + 4 = 4x + 2$$

This is a quadratic equation, so the equivalent equation in standard form is:

$$x^2 - x + 2 = 0$$

This equation can be solved by factoring into:

$$(x - 2)(x + 1) = 0$$

Setting both factors equal to zero results in $x = 2$ and $x = -1$.

32. E: There will be no more coyotes when the population is 0, so set y equal to 0 and solve the quadratic equation:

$$0 = -(x - 2)^2 + 1600$$

Subtract 1600 from both sides, and divide through by -1. This results in:

$$1600 = (x - 2)^2$$

Then, take the square root of both sides. This process results in the following equation:

$$\pm 40 = x - 2$$

Adding 2 to both sides results in two solutions: $x = 42$ and $x = -38$. Because the problem involves years after 2000, the only solution that makes sense is 42. Add 42 to 2000, so therefore in 2042 there will be no more coyotes.

33. B: The ball is back at the starting point when the function is equal to 800 feet. Therefore, this results in solving the equation:

$$800 = -32t^2 + 90t + 800$$

Subtract 800 off of both sides and factor the remaining terms to obtain:

$$0 = 2t(-16 + 45t)$$

Setting both factors equal to 0 result in $t = 0$, which is when the ball was thrown up initially, and:

$$t = \frac{45}{16} = 2.8 \text{ seconds}$$

Therefore, it will take the ball 2.8 seconds to come back down to its staring point.

34. C: To find the zeros, set the function equal to 0 and factor the polynomial. Because there are four terms, it should be factored by grouping. Factor a common factor out of the first set of two terms, and then find a shared binomial factor in the second set of two terms. This results in:

$$x^2(x - 2) - 9(x - 2) = 0$$

The binomial can then be factored out of each set to get $(x^2 - 9)(x - 2) = 0$. This can be factored further as:

$$(x + 3)(x - 3)(x - 2) = 0$$

Setting each factor equal to zero and solving results in the three zeros -3, 3, and 2.

35. B: Given a rational function, the expression in the denominator can never be equal to 0. To find the domain, set the denominator equal to 0 and solve for x. This results in $2 - x = 0$, and its solution is $x = 2$. This value needs to be excluded from the set of all real numbers, and therefore the domain written in interval notation is $(-\infty, 2) \cup (2, \infty)$.

36. C: Plugging the function into the formula results in $\frac{4(x+h)-2-(4x-2)}{h}$, which is simplified to:

$$\frac{4x + 4h - 2 - 4x + 2}{h} = \frac{4h}{h} = h$$

This value is also equal to the derivative of the given function. The derivative of a linear function is its slope.

37. D: The correct matrix to describe the given system of equations is the Choice *D* because it has values that correspond to the coefficients in the right order. The top row corresponds to the coefficient in the first equation, the second row corresponds to the coefficients in the second equation, and the third row corresponds to the coefficients in the third equation. The second matrix (Choice *B*) is filled with the three variables in the system. One thing to also look for is the sign on the numbers, to make sure the signs are correct from the equation to the matrix.

38. C: By switching from a radical expression to rational exponents:

$$\sqrt[4]{x^6} = x^{\frac{6}{4}} = x^{\frac{3}{2}}$$

Also, properties of exponents can be used to simplify $\frac{x}{x^3}$ into:

$$x^{1-3} = x^{-2} = \frac{1}{x^2}$$

The other terms can be left alone, resulting in an equivalent expression:

$$x^{\frac{3}{2}} - \frac{1}{x^2} + x - 2$$

39. B: The slopes of perpendicular lines are negative reciprocals, meaning their product is equal to -1. The slope of the line given needs to be found. Its equivalent form in slope-intercept form is $y = -\frac{4}{7}x + 23$, so its slope is $-\frac{4}{7}$. The negative reciprocal of this number is $\frac{7}{4}$. The only line in the options given with this same slope is:

$$y = \frac{7}{4}x - 12$$

40. D: This system can be solved using substitution. Plug the second equation in for y in the first equation to obtain $2x - 8x = 6$, which simplifies to $-6x = 6$. Divide both sides by 6 to get $x = -1$, which is then back-substituted into either original equation to obtain $y = -8$.

41. B: The outlier is 35. When a small outlier is removed from a data set, the mean and the median increase. The first step in this process is to identify the outlier, which is the number that lies away from the given set. Once the outlier is identified, the mean and median can be recalculated. The mean will be affected because it averages all of the numbers. The median will be affected because it finds the middle number, which is subject to change because a number is lost. The mode will most likely not change because it is the number that occurs the most, which will not be the outlier if there is only one outlier.

42. A: Division can be used to solve this problem. The division necessary is:

$$\frac{5.972 \times 10^{24}}{7.348 \times 10^{22}}$$

To compute this division, divide the constants first then use algebraic laws of exponents to divide the exponential expression. This results in about 0.8127×10^2, which written in scientific notation is 8.127×10^1.

43. D: The graph represents the function x^2 shifted down 4 units, which is:

$$f(x) = x^2 - 4$$

Also, one could plug the ordered pairs given in the graph into each choice. The only one that satisfies (0, -4), (1, -3), (-1, -3), (2,0), and (-2,0) is Choice *D*.

44. C: To find the midpoint, the average of both the x- and y-coordinates must be found. The average of the x-coordinates is:

$$\frac{1 + (-2)}{2} = -\frac{1}{2}$$

The average of the y-coordinates is:

$$\frac{3 + (-5)}{2} = -1$$

Therefore, the midpoint is:

$$\left(-\frac{1}{2}, -1\right)$$

45. B: The area of a trapezoid is equal to the average of its two bases times its height. In this case, the area is:

$$\left(\frac{4+6}{2}\right) \cdot 9 = 5 \cdot 9 = 45 \text{ square centimeters}$$

46. E: The diameter of the circle is equal to two times its radius. The radius is the distance from the center of the circle to the point (4, 5). The distance formula is used to calculate this measurement. Therefore, the radius is:

$$\sqrt{(x_2 - x_1)^2 + (y_2 - y_1)^2}$$

$$\sqrt{(4 - 0)^2 + (5 - 0)^2}$$

$$\sqrt{16 + 25} = \sqrt{41}$$

and the diameter is:

$$2\sqrt{41}$$

47. C: The Pythagorean Theorem can be utilized to find the length of the rectangle when the width and the diagonal are known. Therefore:

$$8^2 + b^2 = 16^2$$

$$b^2 = 256 - 64 = 192$$

This rectangle has a length of:

$$b = \sqrt{192} = 8\sqrt{3} \text{ centimeters}$$

The area of the rectangle is length times width, which is:

$$8 \cdot 8\sqrt{3} = 64\sqrt{3} \text{ square centimeters}$$

48. D: The total of the three angles must equal 180 degrees. Therefore, $15x = 180$, or $x = \frac{180}{15} = 12$. Plugging this value in for x in the expression results in 36, 48, and 96 degrees.

49. D: The exponential rules $(ab)^m = a^m b^m$ and $(a^m)^n = a^{mn}$ can be used to rewrite the expression as:

$$4^4 y^{12} \cdot 3^2 y^{14}$$

The coefficients are multiplied together and the exponential rule $a^m a^n = a^{m+n}$ is then used to obtain the simplified form $2304 y^{26}$.

50. B: Look on the horizontal axis to find 3:00 p.m. Move up from 3:00 p.m. to reach the dot on the graph. Move horizontally to the left to the horizontal axis to between 20 and 25; the best answer choice is 22. The answer of 25 is too high above the projected time on the graph, and the answers of 20 and 16 degrees are too low.

51. B: The number of representatives varies directly with the population, so the equation necessary is $N = k \cdot P$, where N is number of representatives, k is the variation constant, and P is total population in millions. Plugging in the information for New York allows k to be solved for. This process gives $19.8 = k \times 27$, so $k = 0.73$. Therefore, the formula for number of representatives given total population in billions is:

$$N = .73 \times P$$

Plugging in $P = 11.6$ for Ohio results in $N = 8.6$, which rounds up to 9 total Representatives.

52. E: The mean is found by adding all the times together and dividing by the number of times recorded.

$$25 + 18 + 23 + 28 + 30 + 22.5 + 23 + 33 + 20 = 222.5$$

divided by $9 = 24.7$. Rounding to the nearest minute, the mean is 25.

53. C: The mode is the time from the data set that occurs most often. The number 23 occurs twice in the data set, while all others occur only once, so the mode is 23.

54. A: To find the median of a data set, you must first list the numbers from smallest to largest, and then find the number in the middle. If there are two numbers in the middle, as in this data set, add the two numbers in the middle together and divide by 2. Putting this list in order from smallest to greatest yields 18, 20, 22.5, 23, 23, 25, 28, 30, and 33, where 23 is the middle number.

55. B: The perimeter is found by adding the length of all the exterior sides. When the given dimensions are added, the perimeter is 22 meters. The equation to find the perimeter can be:

$$P = 5 + 1.5 + 1.2 + 4.5 + 3.8 + 6 = 22$$

The last two dimensions can be found by subtracting 1.2 from 5, and adding 1.5 and 4.5, respectively.

56. C: First, the parentheses are removed to obtain:

$$\frac{a^5 b^3 c^7}{a^2 b^4 c^6}$$

Then, applying exponential law $\frac{x^m}{x^n} = x^{m-n}$ simplifies the expression into:

$$\frac{a^3 c}{b}$$

57. B: First, plug 500 in for y, which represents her sales. Then, set this expression equal to 200, her total earnings, for:

$$8x + 0.00048(500)^2 = 200$$

Simplifying:

$$8x + 120 = 200$$

Solving this for x gives x = 10, and therefore, she worked 10 hours.

58. A: First, factor the polynomial completely. It factors into:

$$2(3x^2 - 7x - 10)$$

$$2(3x - 10)(x + 1) = 0$$

Then, set each factor equal to 0 and solve for x to obtain the two solutions $\frac{10}{3}$ and -1. The problem specifies that we are looking for a negative solution, so the answer is -1.

59. B: For a system of equations to have no solution, the equations must be parallel. Parallel lines have the same slope and different y-intercepts. The equation of the line shown is:

$$y = -\frac{4}{5}x + 3$$

Therefore, the only line given that is parallel is $y = -\frac{4}{5}x + 16$ because it has the same slope and different y-intercepts.

60. D: Let x represent the first odd integer. The next odd integer would then be $x + 2$, and the third would be $x + 4$. The problem states that four times the first ($4x$) plus 2 times the second ($2(x + 2)$) plus 2 less than the third ($x + 4 - 2$) add up to 111. Therefore:

$$4x + 2(x + 2) + x + 4 - 2 = 111$$

The left side simplifies to $7x + 6$. The solution of $7x + 6 = 111$ is $x = 15$. Therefore, the three consecutive odd integers are 15, 17, and 19.

61. B: Let x be equal to the fifth test score. Therefore, in order to receive, at minimum, a B in the class, the student must have:

$$\frac{74 + 76 + 82 + 84 + x}{5} = 80$$

Therefore:

$$\frac{316 + x}{5} = 80$$

Solving for x gives $316 + x = 400$, or $x = 84$. Therefore, he must receive at least an 84 out of 100 on the fifth test to receive a B in the course.

62. A: Let m represent the number of male students and f represent the number of female students. Therefore:

$$m + f = 900$$

Also, we know that $\frac{m}{f} = \frac{3}{2}$ because the ratio of male to female students is 3/2. Cross-multiplying this proportion results in $2m = 3f$, or $f = \frac{2}{3}f$. Substituting this into the first equation results in $m + \frac{2}{3}m = 900$, so $\frac{5}{3}m = 900$. Solving this for m results in $m = 540$ students.

63. C: There are 23 total balls in the bag. Because 8 of them are orange, 15 of them are not orange. The probability of selecting a ball that is not orange is equal to the number of balls that are not orange over the total number of soccer balls, which can be written as $\frac{15}{23}$.

64. B: The bag contains:

$$\left(\frac{7}{10}\right) \cdot 210 = 147 \text{ red cards}$$

Therefore, the rest of the cards are not red. There are $210 - 147 = 63$ cards that are not red.

Practice Test #2

1. What is the result of dividing 24 by $\frac{8}{5}$?

    a. $\frac{5}{3}$

    b. $\frac{3}{5}$

    c. $\frac{120}{8}$

    d. 15

    e. $\frac{24}{5}$

2. Subtract $\frac{5}{14}$ from $\frac{5}{24}$. Which of the following is the correct result?

    a. $\frac{25}{168}$

    b. 0

    c. $-\frac{25}{168}$

    d. $\frac{1}{10}$

    e. $-\frac{1}{10}$

3. Which of the following is a correct mathematical statement?

    a. $\frac{1}{3} < -\frac{4}{3}$

    b. $-\frac{1}{3} > \frac{4}{3}$

    c. $\frac{1}{3} > \frac{4}{3}$

    d. $-\frac{1}{3} \geq \frac{4}{3}$

    e. $\frac{1}{3} > -\frac{4}{3}$

4. What is the area of the following figure?

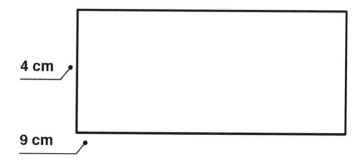

a. 26 cm
b. 36 cm
c. 13 cm²
d. 36 cm²
e. 65 cm²

5. What is the volume of the given figure?

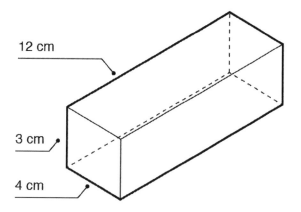

a. 36 cm²
b. 144 cm³
c. 72 cm³
d. 36 cm³
e. 144 cm²

6. What type of units are used to describe surface area?
a. Square
b. Cubic
c. Single
d. Quartic
e. Volumetric

7. What is the solution to the equation $10 - 5x + 2 = 7x + 12 - 12x$?

    a. $x = 12$

    b. $x = 1$

    c. $x = 0$

    d. All real numbers

    e. No solution

8. Gina took an algebra test last Friday. There were 35 questions, and she answered 60% of them correctly. How many correct answers did she have?

    a. 35

    b. 20

    c. 21

    d. 25

    e. 18

9. A car manufacturer usually makes 15,412 SUVs, 25,815 station wagons, 50,412 sedans, 8,123 trucks, and 18,312 hybrids a month. About how many cars are manufactured each month?

    a. 120,000

    b. 200,000

    c. 300,000

    d. 12,000

    e. 20,000

10. Using the graph below, what is the mean number of visitors for the first 4 hours?

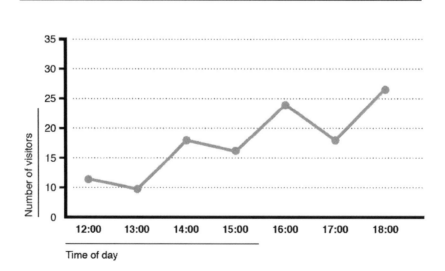

Museum Visitors

Number of visitors

Time of day

    a. 12

    b. 13

    c. 14

    d. 15

    e. 16

11. What type of relationship is there between age and attention span as represented in the graph below?

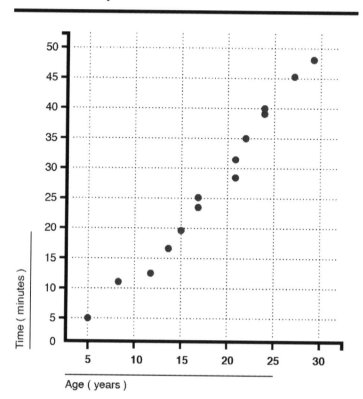

Attention Span

Time (minutes)

Age (years)

a. No correlation
b. Positive correlation
c. Negative correlation
d. Weak correlation
e. Inverse correlation

12. How many kiloliters are in 6 liters?
a. 6,000
b. 600
c. 0.006
d. 0.0006
e. 0.06

13. What is the domain of the logarithmic function $f(x) = \log_2(x - 2)$?
a. 2
b. $(-\infty, \infty)$
c. $(0, \infty)$
d. $(2, \infty)$
e. $(-\infty, 2)$

14. The function $f(t) = \frac{20{,}000}{1+10e^{-2t}}$ represents the number of people who catch a disease t weeks after its initial outbreak in a population of 20,000 people. How many people initially had the disease at the time of the initial outbreak? Round to the nearest whole number.

    a. 20,000

    b. 1,818

    c. 2,000

    d. 0

    e. 18,181

15. How is a transposition of a matrix performed?

    a. Multiply each number by negative 1

    b. Switch the rows and columns

    c. Reverse the order of each row

    d. Find the inverse of each number

    e. Divide the first number in each row by the last number in the last column

16. Given the linear function $g(x) = \frac{1}{4}x - 2$, which domain value corresponds to a range value of $\frac{1}{8}$?

    a. $\frac{17}{2}$

    b. $-\frac{63}{32}$

    c. 0

    d. $\frac{2}{17}$

    e. $\frac{15}{2}$

17. How many possible positive zeros does the polynomial function $f(x) = x^4 - 3x^3 + 2x + x - 3$ have?

    a. 4

    b. 5

    c. 2

    d. 1

    e. 3

18. Which of the following is equivalent to $16^{\frac{1}{4}}16^{\frac{1}{2}}$?

    a. 8

    b. 16

    c. 4

    d. 4,096

    e. 64

19. What is the solution to the following linear inequality?
$$7 - \frac{4}{5}x < \frac{3}{5}$$

    a. $(-\infty, 8)$

    b. $(8, \infty)$

    c. $[8, \infty)$

    d. $(-\infty, 8]$

    e. $(-\infty, \infty)$

20. What is the solution to the following system of linear equations?
$$2x + y = 14$$
$$4x + 2y = -28$$
    a. (0, 0)
    b. (14, -28)
    c. (-14, 28)
    d. All real numbers
    e. There is no solution

21. Triple the difference of five and a number is equal to the sum of that number and 5. What is the number?
    a. 5
    b. 2
    c. 5.5
    d. 2.5
    e. 1

22. Which of the following values could NOT represent a probability?
    a. 0.0123

    b. 0.99999

    c. $\frac{1}{10}$

    d. $\frac{3}{2}$

    e. $\frac{4}{5}$

23. Which of the following is the correct factorization of the polynomial $625x^8 - 25y^4$?
    a. $25(5x^4 + y^2)(5x^2 - y)(x^2 + y)$

    b. $25(5x^4 + y^2)(5x^4 - y^2)$

    c. $(25x^4 + 5y^2)(25x^4 - 5y^2)$

    d. $(25x^4 - y^2)(5x^4 - 5y^2)$

    e. $5(5x^4 - y^2)(5x^4 + y^2)$

24. Which of the following is the result of the expression $\frac{14x}{y^2} - \frac{10y}{x^2}$ evaluated for $x = -2$ and $y = -3$?
    a. $\frac{79}{18}$

    b. $\frac{191}{18}$

    c. $-\frac{191}{18}$

    d. $-\frac{79}{18}$

    e. $\frac{57}{18}$

25. Which of the following is the x-coordinate of the solution of the following system of equations?
$$3x + 2y = 11$$
$$-x + y = 3$$

  a. 4
  b. 1
  c. 3
  d. -1
  e. There is no solution

26. If $4^x = 1024$, what is the value of $9x^4$?
  a. 5
  b. 5,625
  c. 45
  d. 625
  e. 9,216

27. Which of the following represents the equation $z = \frac{xy-4}{x-y}$ solved for y?

  a. $y = \dfrac{xz + 4}{z + x}$

  b. $y = \dfrac{z+x}{xz+4}$

  c. $y = \dfrac{x-y}{xy-4}$

  d. $y = (xz + 4)(z + x)$

  e. $y = \dfrac{xz - 4}{z - x}$

28. What is the solution set of $\sqrt[7]{8x - 4} = 3$?
  a. $\{0, \frac{2191}{8}\}$

  b. $\{0\}$

  c. $\left\{\frac{2191}{8}\right\}$

  d. $\left\{\frac{7}{8}\right\}$

  e. There is no solution

29. If $f(x) = 4\sqrt{x}$ and $f(g(x)) = 4(x + 2)$, which of the following could be $g(x)$?
  a. $x + 2$
  b. $4x + 4$
  c. $\sqrt{x + 2}$
  d. $(x + 2)^2$
  e. $x^2 + 4$

30. The following represents the graph of the function $f(x)$. How many real solutions does $f(x) = 0$ have?

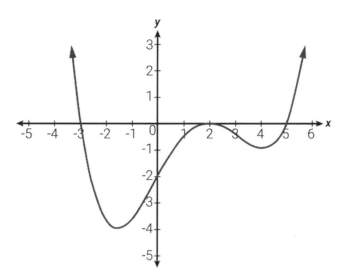

a. 0
b. 1
c. 2
d. 3
e. 4

31. A polynomial function is defined by $f(x) = 3x^3 + 2x^2 + 9x - ax + 3 - 2a$. If $f(2) = 0$, what is a?

a. 53

b. 0

c. $\frac{57}{4}$

d. $\frac{43}{4}$

e. $\frac{53}{4}$

125

32. The graph below shows the radical function $f(x) = \sqrt{x}$.

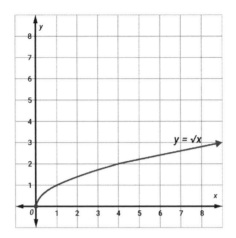

What transformations would be involved if $g(x) = \sqrt{x - 4} + 3$ were to be graphed?
- a. Up 3, right 4
- b. Down 3, left 4
- c. Up 3, left 4
- d. Up 4, right 3
- e. Down 4, left 3

33. If $\log_6 \frac{1}{1296} = x$, what is the value of x?
- a. -2
- b. -3
- c. -4
- d. 3
- e. 4

34. Which of the following is equivalent to the logarithmic function $f(x) = 3\log_7 x - \frac{1}{2}\log_7 y + 2\log_7 z$?
- a. $f(x) = \log_7 x^3 yz^2$
- b. $f(x) = \log_7 \frac{x^3 \sqrt{y} z^2}{y}$
- c. $f(x) = \log_7(3x - .5y + 2z)$
- d. $f(x) = \log_7 3xyz$
- e. $f(x) = \log_7 \frac{x^2 y^2}{z}$

35. The graph of a polynomial crosses the y-axis 4 times. Which of the following could represent the formula for the function? Let *a, b, c,* and *d* represent real-numbered constants.
- a. $f(x) = x^3 - 3x + b$
- b. $g(x) = x^2 + 4x + a$
- c. $h(x) = x^4 + 4$
- d. $g(x) = x^4 + bx^3 + cx - d$
- e. $f(x) = mx + b$

36. The graph of a function $h(x)$ is found by shifting the graph of $g(x)$ down 7 units, to the left 8 units, and finally, up 2 units. If $g(x) = \sqrt{x-2}$, what is $h(x)$?

    a. $h(x) = \sqrt{x-10} + 5$
    b. $h(x) = \sqrt{x+6} + 5$
    c. $h(x) = \sqrt{x-6} - 5$
    d. $h(x) = \sqrt{x+6} - 5$
    e. $h(x) = \sqrt{x-10} - 5$

37. Given the graph of the following piecewise defined function $f(x)$, what is its domain?

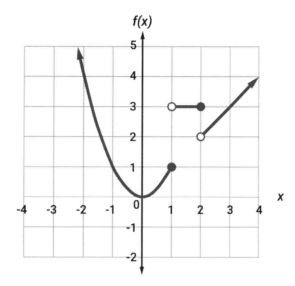

    a. $(-\infty, 1) \cup (1, \infty)$
    b. $(-\infty, \infty)$
    c. $(-\infty, 1) \cup (1,2) \cup (2, \infty)$
    d. $(0, \infty)$
    e. $[0, \infty)$

38. What is the period of the function $f(x) = 4\sec(2x) + 1$?

    a. 2π
    b. $\frac{\pi}{2}$
    c. π
    d. 4
    e. 2

39. Each year, a family goes to the grocery store every week and spends $105. About how much does the family spend annually on groceries?

    a. $10,000
    b. $50,000
    c. $500
    d. $5,000
    e. $1,200

40. Bindee is having a barbeque on Sunday and needs 12 packets of ketchup for every 5 guests. If 60 guests are coming, how many packets of ketchup should she buy?

    a. 100
    b. 12
    c. 144
    d. 60
    e. 300

41. If x is directly proportional to y and if $x = 8$ when $y = \frac{1}{12}$, then which of the following describes an equation that relates x and y?

    a. $x = 96y$
    b. $y = 96x$
    c. $x = 12y$
    d. $y = 1.5x$
    e. $x = 1.5y$

42. If the ratio of x to y is 1:8, what is the product of x and y when $y = 48$?

    a. 96
    b. 183
    c. 48
    d. 288
    e. 144

43. The percent increase from 8 to 18 is equivalent to the percent increase from 234 to what number?

    a. 468
    b. 526
    c. 526.5
    d. 125
    e. 527

44. On the first four tests this semester, a student received the following scores out of 100: 74, 76, 82, and 84. What score must that student receive on the fifth test to receive a B in the class? Assume that the final test is also out of 100 points and that to receive a B in the class, he must have at least an 80% average.

    a. 80
    b. 84
    c. 76
    d. 78
    e. 82

45. In a class of 3 girls and 5 boys, the average score for boys was an 82 and the average for girls was 84. What was the class average?

    a. 82
    b. 82.75
    c. 83
    d. 83.75
    e. 84

46. How many 5-letter orderings of the word FRANCHISE can be made if no letters are repeated?
- a. 3,024
- b. 504
- c. 72
- d. 15,120
- e. 45

47. A set of cards contains n numbers, one of which is an odd number. If one card is randomly selected from the set, what is the probability that the card is even?
- a. $\frac{1}{n}$
- b. $\frac{1}{n-1}$
- c. $\frac{n-2}{n-1}$
- d. $n-1$
- e. $\frac{n-1}{n}$

48. At a school raffle, 1000 tickets are sold for $1 each with the possibility of winning a $200 or a $50 prize. What is the expected value of the gain?
- a. $0.75
- b. $1
- c. -$0.75
- d. $1.25
- e. $0.95

49. Which of the following pairs of angles could NOT be the smaller and larger interior angles of a parallelogram?
- a. 120°, 60°
- b. 135°, 55°
- c. 110°, 60°
- d. 20°, 160°
- e. 40°, 140°

50. The sides of a triangle have the following lengths: 4 inches, 4 inches, and 28 inches. If the smallest angle within the triangle has a measurement of 20°, what is the measure of the largest angle?
- a. 100°
- b. 140°
- c. 120°
- d. 160°
- e. 90°

51. In the following triangle, what is the value of $\sin x$?

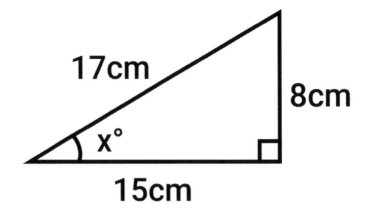

a. $\frac{8}{17}$

b. $\frac{15}{17}$

c. $\frac{8}{15}$

d. $\frac{15}{8}$

e. $\frac{x}{8}$

52. What is the area of the following parallelogram?

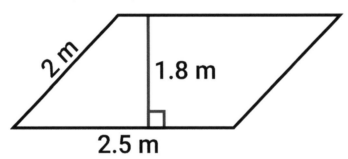

    a. 5 square meters
    b. 3.6 square meters
    c. 4 square meters
    d. 4.5 square meters
    e. 3.24 square meters

53. In $\triangle ABC$, $\angle B$ measures 60% and $\angle A$ is a right triangle. If $\overline{BC}$ is 16 centimeters long, what is the area of the triangle?
    a. 32 square centimeters
    b. $32\sqrt{3}$ square centimeters
    c. 64 square centimeters
    d. $8\sqrt{3}$ square centimeters
    e. $64\sqrt{3}$ square centimeters

54. A soup company fills their soup cans 90% of the way to the top to eliminate spilling when consumers open them up. Each cylindrical can contains 16 ounces of soup. The containers are 6 inches tall and 3 inches wide. Approximately how many cubic inches of space does one ounce of soup take up?
   a. 2.4
   b. 2.1
   c. 3.2
   d. 3.5
   e. 4.0

55. What is the equation of the line that passes through the two points (-3, 7) and (-1, -5)?
   a. $y = 6x + 11$
   b. $y = 6x$
   c. $y = -6x - 11$
   d. $y = -6x$
   e. $y = 6x - 11$

56. The percentage of smokers above the age of 18 in 2000 was 23.2 percent. The percentage of smokers above the age of 18 in 2015 was 15.1 percent. Find the average rate of change in the percent of smokers above the age of 18 from 2000 to 2015.
   a. -.54 percent
   b. -54 percent
   c. -5.4 percent
   d. -15 percent
   e. -1.5 percent

57. A study of adult drivers finds that it is likely that an adult driver wears his seatbelt. Which of the following could be the probability that an adult driver wears his seat belt?
   a. 0.90
   b. 0.05
   c. 0.25
   d. 0
   e. 1.5

58. In order to estimate deer population in a forest, biologists obtained a sample of deer in that forest and tagged each one of them. The sample had 300 deer in total. They returned a week later and harmlessly captured 400 deer, and 5 were tagged. Use this information to estimate how many total deer were in the forest.
   a. 24,000 deer
   b. 30,000 deer
   c. 40,000 deer
   d. 100,000 deer
   e. 120,000 deer

59. Which of the following is the equation of a vertical line that runs through the point (1, 4)?
   a. $x = 1$
   b. $y = 1$
   c. $x = 4$
   d. $y = 4$
   e. $x = y$

60. What is the missing length *x*?

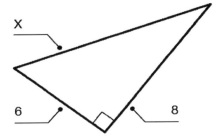

    a. 6
    b. -10
    c. 10
    d. 100
    e. 14

61. At a printing press, 510,000 sheets of paper are required to make 600 phone directories. How many sheets of paper are required to make 750 phone directories?
    a. 615,000
    b. 820,000
    c. 757,500
    d. 637,500
    e. 910,500

62. A number is randomly selected from all integers greater than 0 and less than 27. What is the probability that the randomly selected number is prime?
    a. $\frac{8}{13}$

    b. $\frac{5}{26}$

    c. $\frac{9}{26}$

    d. $\frac{1}{2}$

    e. $\frac{5}{13}$

63. The average of a set of 9 numbers is 10. If 6 is added to each number in the original set of numbers, what is the average of this new set of numbers?
    a. 11
    b. 12
    c. 14
    d. 16
    e. 20

64. What is the correct factorization of the following binomial?
$$2y^3 - 128$$

    a. $2(y + 8)(y - 8)$
    b. $2(y - 4)(y^2 + 4y + 16)$
    c. $2(y + 4)(y - 4)^2$
    d. $2(y - 4)^3$
    e. $2(y - 4)(y^2 + 4y + 16)$

Answer Explanations #2

1. D: Division is completed by multiplying times the reciprocal. Therefore:

$$24 \div \frac{8}{5}$$

$$\frac{24}{1} \times \frac{5}{8}$$

$$\frac{3 \times 8}{1} \times \frac{5}{8}$$

$$\frac{15}{1} = 15$$

2. C: Common denominators must be used. The LCD is 168, so each fraction must be converted to have 168 as the denominator.

$$\frac{5}{24} - \frac{5}{14}$$

$$\frac{5}{24} \times \frac{7}{7} - \frac{5}{14} \times \frac{12}{12}$$

$$\frac{35}{168} - \frac{60}{168} = -\frac{25}{168}$$

3. E: The correct mathematical statement is the one in which the number to the left on the number line is less than the number to the right on the number line. It is written in answer E that $\frac{1}{3} > -\frac{4}{3}$, which is the same as $-\frac{4}{3} < \frac{1}{3}$, a correct statement.

4. D: The area for a rectangle is found by multiplying the length by the width. The area is also measured in square units, so the correct answer is Choice D. The answer of 26 is the perimeter. The answer of 13 is found by adding the two dimensions instead of multiplying.

5. B: The volume of a rectangular prism is found by multiplying the length by the width by the height. This formula yields an answer of 144 cubic units. The answer must be in cubic units because volume involves all three dimensions. Each of the other answers have only two dimensions that are multiplied, and one dimension is forgotten, as in D, where 12 and 3 are multiplied, or have incorrect units, as in E.

6. A: Surface area is a type of area, which means it is measured in square units. Cubic units are used to describe volume, which has three dimensions multiplied by one another. Quartic units describe measurements multiplied in four dimensions.

7. D: First, like terms are collected to obtain $12 - 5x = -5x + 12$. Then, if the addition principle is used to move the terms with the variable, $5x$ is added to both sides and the mathematical statement $12 = 12$ is obtained. This is always true; therefore, all real numbers satisfy the original equation.

8. C: Gina answered 60% of 35 questions correctly; 60% can be expressed as the decimal 0.60. Therefore, she answered $0.60 \times 35 = 21$ questions correctly.

9. A: Rounding can be used to find the best approximation. All of the values can be rounded to the nearest thousand. 15,412 SUVs can be rounded to 15,000. 25,815 station wagons can be rounded to 26,000. 50,412 sedans can be rounded to 50,000. 8,123 trucks can be rounded to 8,000. Finally, 18,312 hybrids can be rounded to 18,000. The sum of the rounded values is 117,000, which is closest to 120,000.

10. C: The mean for the number of visitors during the first 4 hours is 14. The mean is found by calculating the average for the four hours. Adding up the total number of visitors during those hours gives:

$$12 + 10 + 18 + 16 = 56$$

Then:

$$56 \div 4 = 14$$

11. B: The relationship between age and time for attention span is a positive correlation because the general trend for the data is up and to the right. As the age increases, so does attention span.

12. C: There are 0.006 kiloliters in 6 liters because 1 liter=0.001kiloliters. The conversion comes from the chart where the prefix kilo is found three places to the left of the base unit.

13. D: The argument of a logarithmic function has to be greater than or equal to zero. Basically, one cannot take the logarithm of a negative number or 0. Therefore, to find the domain, set the argument greater than 0 and solve the inequality. This results in $x - 2 > 0$, or $x > 2$. Therefore, in order to obtain an output of the function, the number plugged into the function must be greater than 2. This domain is represented as $(2, \infty)$.

14. B: The time of the initial outbreak corresponds to $t = 0$. Therefore, 0 must be plugged into the function. This results in:

$$\frac{20,000}{1 + 10e^0} = \frac{20,000}{1 + 10} = \frac{20,000}{11} = 1,818.182$$

which rounds to 1,818. Therefore, there were 1,818 people in the population that initially had the disease.

15. B: The correct choice is *B* because the definition of transposing a matrix says that the rows and columns should be switched. For example, the matrix:

$$\begin{bmatrix} 3 & 4 \\ 2 & 5 \\ 1 & 6 \end{bmatrix}$$

can be transposed into:

$$\begin{bmatrix} 3 & 2 & 1 \\ 4 & 5 & 6 \end{bmatrix}$$

Notice that the first row, 3 and 4, becomes the first column. The second row, 2 and 5, becomes the second column. This is an example of transposing a matrix.

16. A: The range value is given, and this is the output of the function. Therefore, the function must be set equal to $\frac{1}{8}$ and solved for x. Thus, $\frac{1}{8} = \frac{1}{4}x - 2$ needs to be solved. The fractions can be cleared by multiplying times the LCD 8. This results in:

$$1 = 2x - 16$$

Add 16 to both sides and divide by 2 to obtain $x = \frac{17}{2}$.

17. E: Using Descartes' Rule of Signs, count the number of sign changes in coefficients in the polynomial. This results in the number of possible positive zeros. The coefficients are 1, -3, 2, 1, and -3, so the sign changes from 1 to -3, -3 to 2, and 1 to -3, a total of 3 times. Therefore, there are at most 3 positive zeros.

18. A: The corresponding expression written using common denominators of the exponents is $16^{\frac{1}{4}}16^{\frac{2}{4}}$, and then the expression is written as:

$$(16 \cdot 16^2)^{\frac{1}{4}}$$

This can be written in radical notation as:

$$\sqrt[4]{16^3} = \sqrt[4]{4{,}096} = 8$$

19. B: The goal is to first isolate the variable. The fractions can easily be cleared by multiplying the entire inequality by 5, resulting in:

$$35 - 4x < 3$$

Then, subtract 35 from both sides and divide by -4. This results in $x > 8$. Notice the inequality symbol has been flipped because both sides were divided by a negative number. The solution set, all real numbers greater than 8, is written in interval notation as $(8, \infty)$. A parenthesis shows that 8 is not included in the solution set.

20. E: This system can be solved using the method of substitution. Solving the first equation for y results in:

$$y = 14 - 2x$$

Plugging this into the second equation gives $4x + 2(14 - 2x) = -28$, which simplifies to $28 = -28$, an untrue statement. Therefore, this system has no solution because no x value will satisfy the system.

21. D: Let x be the missing quantity. The problem can be expressed as the following equation:

$$3(5 - x) = x + 5$$

Distributing the 3 results in:

$$15 - 3x = x + 5$$

Subtract 5 from both sides, add $3x$ to both sides, and then divide both sides by 4. This results in:

$$\frac{10}{4} = \frac{5}{2} = 2.5$$

22. D: A probability cannot be less than 0 or exceed 1. Because 3/2 is equal to 1.5, this value cannot represent a probability. This value would mean something happening 150% of the time, which does not make sense.

23. B: First factor out the common factor of 25, resulting in:

$$25(25x^8 - y^4)$$

The resulting polynomial in parentheses is a difference of squares, and therefore, the entire polynomial factors into:

$$25(5x^4 + y^2)(5x^4 - y^2)$$

This can be checked by multiplication.

24. A: First, substitute -2 for x and -3 for y. This results in:

$$\frac{14(-2)}{(-3)^2} - \frac{10(-3)}{(-2)^2}$$

Multiply the numerators and evaluate the exponents in the denominators to obtain:

$$-\frac{28}{9} - \left(\frac{-30}{4}\right) = -\frac{28}{9} + \frac{15}{2}$$

The common denominator is 18, and therefore, this expression is equivalent to:

$$-\frac{56}{18} + \frac{135}{18} = \frac{79}{18}$$

25. B: This system can be solved using substitution. First, solve the second equation for y, resulting in:

$$y = x + 3$$

This equation then gets substituted in for y in the first equation to obtain:

$$3x + 2(x + 3) = 11$$

Solving this equation for x gives $3x + 2x + 6 = 11$, so $5x + 6 = 11$, or $5x = 5$. Dividing both sides by 5 results in $x = 1$.

26. B: First, solve for x in the given equation. Because $4^5 = 1025$, we know that $x = 5$. Then, plug 5 in for x in the given expression. Following order of operations:

$$9 \cdot 5^4 = 9 \cdot 625 = 5{,}625$$

27. A: First, multiply both sides of the equation times the quantity $(x - y)$. This results in:

$$z(x - y) = xy - 4$$

The goal is to get all y terms isolated onto one side of the equals sign. Distributing results in:

$$zx - zy = xy - 4$$

Moving the y terms to the right side and the 4 to the left side results in:

$$zx + 4 = xy + zy$$

Factoring out a y on the right side gives:

$$zx + 4 = y(x + z)$$

Finally, dividing both sides by the quantity $(x + z)$ gives:

$$y = \frac{zx + 4}{x + z}$$

28. C: To solve the equation, first raise both sides to the 7th power. Therefore:

$$8x - 4 = 3^7 = 2187$$

Add 4 to both sides and divide by 3 to obtain the single solution:

$$x = \frac{2191}{8}$$

29. D: $f(g(x))$ is a composite function where $f(x) = 4\sqrt{x}$ is the outer function. Plugging a function $g(x)$ in for x results in $4(x + 2)$. The only possibility listed that works is $(x + 2)^2$ because:

$$4\sqrt{(x + 2)^2} = 4(x + 2)$$

30. D: The equation has real solutions where the function either touches or crosses the x-axis. This occurs at $x = $ -3, 2, and 5. Therefore, the equation has three real solutions.

31. E: First, plug in 2 for x to obtain:

$$f(2) = 3(2)^3 + 2(2)^2 + 9(2) - a(2) + 3 - 2a$$

$$24 + 8 + 18 + 3 - 4a = 53 - 4a$$

This expression equals 0, so solve $53 - 4a = 0$ for a to obtain $a = \frac{53}{4}$.

32. A: First, the addition of the 3 involves shifting the graph up 3 units. Then, the subtraction of 4 within the radical involves shifting the graph to the right 4 units.

33. C: Write this logarithm in its corresponding exponential form of:

$$6^x = \frac{1}{1296}$$

It is true that $6^4 = 1296$. Therefore, $6^{-4} = \frac{1}{1296}$ and $x = -4$.

34. B: First, the constants in front of each logarithm can be placed as an exponent inside the arguments of the functions as the following:

$$f(x) = \log_7 x^3 - \log_7 y^{\frac{1}{2}} + \log_7 z^2$$

The ½ power is the same as the square root of y, and by using properties of logarithms, we have $f(x) = \log_7 \frac{x^3 z^2}{\sqrt{y}}$, which is equivalent to:

$$f(x) = \log_7 \frac{x^3 \sqrt{y} z^2}{y}$$

35. D: Because there are 4 zeros, the function must be of at least degree 4. Choice *C* is not the answer because the graph of that function never crosses the *y*-axis. Choice *D* is on the only possibility.

36. D: First, function g is shifted down 7 units, resulting in:

$$h(x) = \sqrt{x - 2} - 7$$

Then, shifting it to the left 8 units results in:

$$h(x) = \sqrt{x - 2 + 8} - 7 = \sqrt{x + 6} - 7$$

Finally, shifting the graph up 2 units results in:

$$h(x) = \sqrt{x + 6} - 5$$

37. B: The domain is the reflection of the function onto the x-axis. It contains all x-coordinates that the function covers. Even though the function jumps at $x = 1$ and $x = 2$, the domain is all real numbers.

38. C: The domain of the secant function of the form $f(x) = Asec(Bx) + C$ is $\frac{2\pi}{B}$. In this case, $B = 2$.

39. D: There are 52 weeks in a year, and if the family spends \$105 each week, that amount is close to \$100. A good approximation is \$100 a week for 50 weeks, which is found through the product $50 \times 100 = $ \$5,000.

40. C: This problem involves ratios and percentages. If 12 packets are needed for every 5 people, this statement is equivalent to the ratio $\frac{12}{5}$. The unknown amount x is the number of ketchup packets needed for 60 people. The proportion $\frac{12}{5} = \frac{x}{60}$ must be solved. Cross-multiply to obtain $12 \times 60 = 5x$. Therefore, $720 = 5x$. Divide each side by 5 to obtain $x = 144$.

41. A: Because x is directly proportional to y, $x = ky$, for some constant k. Plugging in what is given yields $8 = k\left(\frac{1}{12}\right)$, which means that $k = 96$. Therefore, $x = 96y$.

42. D: The ratio gives the proportion $\frac{x}{y} = \frac{1}{8}$. If $y = 48$, then $\frac{x}{48} = \frac{1}{8}$ means that $x = 6$. The product of x and y is therefore $(6)(48) = 288$.

43. B: First, we multiply each score times its weight (the number of boys and the number of girls, respectively). Therefore, $3 \cdot 84 = 252$ and $5 \cdot 82 = 410$. The total weighted score is $252 + 410 = 662$. The total weight is 8 because there are eight students. Therefore, divide 662 by 8 to obtain a class average of 82.75.

44. D: The sum of the first 9 numbers is 90, because its average is 10. If 6 is added to each original number, this adds $9 \cdot 6 = 54$ to the original 9 numbers. Therefore, the sum of the new set of numbers is $90 + 54 = 144$. Therefore, the new average is:

$$\frac{144}{9} = 16$$

45. D: Because there are more women than men, we cannot simply find the average of 24 and 27. In this class, 70% are women and 30% are men. A possibility is that the class has 7 women and 3 men. If this is the case, the total ages of women is $7(24) = 168$ and the total ages of men is $3(27) = 81$. The sum of the total ages is:

$$168 + 81 = 249$$

Divide this by 10 to obtain the class average of 24.9.

46. D: The given word has 9 letters. Therefore, in a 5-letter ordering, there are 9 choices for the first letter. Because there are no repeats allowed, there are 8 choices for the second letter, 7 choices for the third letter, 6 choices for the fourth letter, and 5 choices for the fifth letter. Multiplying these values together results in:

$$(9)(8)(7)(6)(5) = 15{,}120 \; possible \; orderings$$

47. E: There are n total cards, which means that the denominator needs to be n (the total number of outcomes). If there is only 1 odd number, then the rest are even. There are n - 1 even cards. Therefore, the probability of selecting an even card is $\frac{n-1}{n}$.

48. C: Because there is one $200 winner and each ticket costs $1, the probability of winning $199 is 1/1000. Also, because there is one $50 winner, the probability of winning $49 is 1/1000. The rest of the outcomes would be losing $1, which has a probability of 998/1000. The expected value is the sum of the products of each outcome and their probabilities. Therefore,

$$\text{Expected value} = \$199 \left(\frac{1}{1000}\right) + \$49 \left(\frac{1}{1000}\right) - 1 \left(\frac{998}{1000}\right)$$

$$-\frac{750}{1000} = -\$0.75$$

Expect to lose $0.75 for every ticket bought.

49. C: The smaller and larger interior angles in a parallelogram must add up to 180°. This pair is the only duo that does not have a sum of 180.

50. B: Because two sides are equal, this is an isosceles triangle. The smallest sides correspond to the 20° angles. The third angle has a measure of:

$$180 - 20 - 20 = 140°$$

51. A: The sine of x is the ratio of the lengths of the side opposite the angle x and the hypotenuse. Those sides are lengths 8 and 17, respectively.

52. D: The area of a parallelogram is equal to base times height. In this figure, the height is the dimension that is 1.8 m long and its base is 2.5 m long. Therefore, the area of this figure is 2.5(1.8) = 4.5 square meters.

53. B: This is a 30-60-90 triangle with hypotenuse $\overline{BC} = 16 \; cm$. Such a special triangle has sides with ratio $x : \sqrt{3}x : 2x$. In this case, $2x = 16$, so $x = 8$ cm. The triangle has sides 8, $8\sqrt{3}$, and 16 cm, with the first two being its base and height. The area of this height is:

$$A = \frac{1}{2}bh$$

$$\tfrac{1}{2}(8)(8\sqrt{3}) = 32\sqrt{3} \text{ square centimeters}$$

54. A: Volume of a cylinder is:

$$V = \pi r^2 h$$

In this case, the radius is half of the width, so the radius is 1.5 in. Therefore:

$$V = \pi(1.5)^2(6) = 13.5\pi \text{ cubic inches}$$

The can is filled 90% of the way up with soup, so the soup takes up:

$$(.9)(13.5\pi) = 12.15\pi \approx 38.17 \text{ cubic inches}$$

Divide this by 16 to obtain the space each ounce of soup takes up. Therefore:

$$\frac{38.17}{16} \approx 2.4 \text{ cubic inches}$$

55. C: First, the slope of the line must be found. This is equal to the change in *y* over the change in *x*, given the two points. Therefore, the slope is -6. The slope and one of the points are then plugged into the slope-intercept form of a line:

$$y - y_1 = m(x - x_1)$$

This results in:

$$y - 7 = -6(x + 3)$$

The -6 is simplified and the equation is solved for y to obtain:

$$y = -6x - 11$$

56. A: The formula for the rate of change is the same as slope: change in *y* over change in *x*. The *y*-value in this case is percentage of smokers and the *x*-value is year. The change in percentage of smokers from 2000 to 2015 was 8.1 percent. The change in x was:

$$2000\text{-}2015 = \text{-}15$$

Therefore:

$$8.1\%/_{-15} = -0.54\%$$

The percentage of smokers decreased 0.54 percent each year.

57. A: The probability of .9 is closer to 1 than any of the other answers. The closer a probability is to 1, the greater the likelihood that the event will occur. The probability of 0.05 shows that it is very unlikely that an adult driver will wear their seatbelt because it is close to zero. A zero probability means that it will not occur. The probability of 0.25 is closer to zero than to one, so it shows that it is unlikely an adult will wear their seatbelt. Choice *E* is wrong because probability must fall between 0 and 1.

58. A: A proportion should be used to solve this problem. The ratio of tagged to total deer in each instance is set equal, and the unknown quantity is a variable *x*. The proportion is:

$$\frac{300}{x} = \frac{5}{400}$$

Cross-multiplying gives 120,000 = 5x, and dividing through by 5 results in 24,000.

59. A: A vertical line has the same x value for any point on the line. Other points on the line would be (1, 3), (1, 5), (1, 9,) etc. Mathematically, this is written as $x = 1$. A vertical line is always of the form $x = a$ for some constant a.

60. C: The Pythagorean Theorem can be used to find the missing length x because it is a right triangle. The theorem states that $6^2 + 8^2 = x^2$, which simplifies into $100 = x^2$. Taking the positive square root of both sides results in the missing value $x = 10$.

61. D: A proportion can be used to solve this problem. Begin with:

$$\frac{600}{510000} = \frac{750}{x}$$

Cross-multiplying results in:

$$600x = 382{,}500{,}000$$

Dividing both sides by 600 results in $x = 637{,}500$, which represents the number of sheets of paper required for 750 phone directories.

62. E: There are 26 integers greater than 0 and less than 27 (the integers from 1 to 26). Out of those integers, the following are prime: 1, 2, 3, 5, 7, 11, 13, 17, 19, and 23. Therefore, 10 out of 26 are prime, which means that the probability of selecting a prime number is:

$$\frac{10}{26} = \frac{5}{13}$$

63. C: First, calculate the percent increase from 8 to 18 as:

$$\frac{18 - 8}{8} = 1.25 = 125\%$$

Then add 125% of 234 onto 234 to obtain:

$$292.5 + 234 = 526.5$$

64. E: First, the common factor 2 can be factored out of both terms, resulting in $2(y^3 - 64)$. The resulting binomial is a difference of cubes that can be factored using the rule $a^3 - b^3 = (a - b)(a^2 + ab + b^2)$ with a = y and b = 4. Therefore, the result is:

$$2(y - 4)(y^2 + 4y + 16)$$

Greetings!

First, we would like to give a huge "thank you" for choosing us and this study guide for your ACT exam. We hope that it will lead you to success on this exam and for your years to come.

Our team has tried to make your preparations as thorough as possible by covering all of the topics you should be expected to know. In addition, our writers attempted to create practice questions identical to what you will see on the day of your actual test. We have also included many test-taking strategies to help you learn the material, maintain the knowledge, and take the test with confidence.

We strive for excellence in our products, and if you have any comments or concerns over the quality of something in this study guide, please send us an email so that we may improve.

As you continue forward in life, we would like to remain alongside you with other books and study guides in our library, such as;

ACCUPLACER: amazon.com/dp/1628455160

SAT: amazon.com/dp/1628455381

We are continually producing and updating study guides in several different subjects. If you are looking for something in particular, all of our products are available on Amazon. You may also send us an email!

Sincerely,
APEX Test Prep
info@apexprep.com

Free Study Tips DVD

In addition to the tips and content in this guide, we have created a FREE DVD with helpful study tips to further assist your exam preparation. **This FREE Study Tips DVD provides you with top-notch tips to conquer your exam and reach your goals.**

Our simple request in exchange for the strategy-packed DVD is that you email us your feedback about our study guide. We would love to hear what you thought about the guide, and we welcome any and all feedback—positive, negative, or neutral. It is our #1 goal to provide you with top quality products and customer service.

To receive your **FREE Study Tips DVD**, email freedvd@apexprep.com. Please put "FREE DVD" in the subject line and put the following in the email:

      a. The name of the study guide you purchased.

      b. Your rating of the study guide on a scale of 1-5, with 5 being the highest score.

      c. Any thoughts or feedback about your study guide.

      d. Your first and last name and your mailing address, so we know where to send your free DVD!

Thank you!

38212606R00083

Made in the USA
Lexington, KY
05 May 2019